Building Resilience to Climate Change in Selected Districts in Zimbabwe

Lindah Mhlanga • Kefasi Nyikahadzoi
Jayne Stack
Editors

Building Resilience to Climate Change in Selected Districts in Zimbabwe

Experiences from the Zimbabwe Resilience Building Fund Programme

Springer

Editors
Lindah Mhlanga
Environment, Climate and Sustainable Development Institute
University of Zimbabwe
Harare, Zimbabwe

Kefasi Nyikahadzoi
Environment, Climate and Sustainable Development Institute
University of Zimbabwe
Harare, Zimbabwe

Jayne Stack
Resilience Knowledge Hub
Zimbabwe Resilience Building Fund
Harare, Zimbabwe

ISBN 978-3-031-80528-8 ISBN 978-3-031-80529-5 (eBook)
https://doi.org/10.1007/978-3-031-80529-5

This Springer imprint is published by the registered company Springer Nature Switzerland AG
The registered company address is: Gewerbestrasse 11, 6330 Cham, Switzerland

Acknowledgements

The authors would like to acknowledge the Resilience Knowledge Hub under the Zimbabwe Resilience Building Fund for availing the Outcome Monitoring data used in writing this book.

About the Book

This book brings together experiences and lessons learnt during the Zimbabwe Resilience Building Programme in 18 districts in Zimbabwe. Zimbabwe, located in the southern Africa region, has been experiencing climate change and variability which has significantly affected rural livelihoods of people dependent entirely on dryland agriculture and natural resources harvesting. This book will be of great interest to scholars, students, policymakers and people involved in building community resilience against climate-related shocks.

Contents

About the Editors and Contributors

Sandra Bhatasara is a sociologist and senior research fellow at the Environment, Climate and Sustainable Development Institute, University of Zimbabwe. Her research focuses on gender studies, livelihoods and agrarian issues, natural resources, climate change adaptation and resilience.

Lazarus Chapungu researches on contemporary environmental issues, climate change, tourism, sustainable development and biodiversity management. He is currently a postdoctoral research fellow at the University of South Africa. He has worked as a senior lecturer in the Department of Physics, Geography and Environmental Science at Great Zimbabwe University. He served as a chapter scientist, contributing author and glossary editor in the 6th Assessment Report (AR6) of the Intergovernmental Panel on Climate Change (IPCC), Working Group III Chap. 2. Lazarus holds the portfolio of Vice Chair of the programmes committee of the board of directors at the Zimbabwe Council of Churches (ZCC), a non-governmental organization seeking to build community resilience to climate change and enable recovery from man-made and natural disasters. His qualifications include a PhD in Environmental Science, a Master's degree in Environmental Policy and Planning, a degree in Geography, a professional diploma in Applied Monitoring and Evaluation and a program certificate in Project Management.

Everson Dahwa is a livestock and wildlife ecologist with over 12 years of experience in research and development. Currently, he is a lecturer in the Department of Livestock, Wildlife and Fisheries at Great Zimbabwe University. His research interests are in developing innovative products and services that promote livelihoods.

Modester Dengedza is a lecturer at the University of Zimbabwe. She is working towards her PhD in Economics. She holds an MSc and BSc in Economics (UZ). She specializes in applied economics. Her research interests are in development economics, agriculture economics, climate change issues and gender. She has done consultancies on climate change and food security issues and gender issues.

Lawrence Dhliwayo is a highly talented and dedicated statistics lecturer at the University of Zimbabwe with more than 10 years of experience. His research area includes time series analysis, econometrics, statistical data analysis, sampling techniques and research methodology. He has published in fractionally integrated times series, time series with conditional heteroscedasticity, time series with seasonality intervention analysis and structural change in time series.

Praise Karuma is a lecturer in the Department of Community and Social Development at the University of Zimbabwe. He has a background in sociology and social anthropology. His MSc thesis focused on women and climate change on a post-fast track farm in peri-urban Harare. He is also a PhD student with his thesis focusing on Gender and Socio-Ecological Resilience to Climate Change.

Samuel Kusangaya is a physical geographer who specializes in hydrology and geo-spatial science applications in hydrology and water resources management and has extensive experience spanning over 20 years. He is a lecturer in the Department of Geography, Geospatial Sciences and Earth Observation at the University of Zimbabwe.

Pamela Madududu is a PhD Agricultural Economics candidate in the Department of Agricultural Economics, Extension and Rural Development at the University of Pretoria. Her research focuses on plant-based milk's socioeconomic and environmental footprints in Sub-Saharan Africa. She lectures at Manicaland State University of Applied Sciences, Zimbabwe. She completed research internships at the World Agroforestry Center, the African Capacity Building Foundation and the Centre for Agriculture and Food Policy. She is also a committee member of the Association of Agricultural Economists in Zimbabwe.

Jeanette Manjengwa is an associate professor who completed her PhD at the Institute for Development Policy and Management (IDPM), University of Manchester, UK; a Master's degree in Environmental Policy and Planning from the University of Zimbabwe; and a BSc (Hons) in Environmental Science from the New University of Ulster, UK. At the University of Zimbabwe, she lectured in the Centre for Applied Social Sciences and was the Deputy Director, at the Institute of Environmental Studies. Professor Manjengwa's main research areas are sustainable and pro-poor development; resilience knowledge and environmental and equity aspects of land; and agrarian reform. Professor Manjengwa was guest editor of the *Development Southern Africa* (2016) Special Issue: 'Towards Resilient Urban Communities'.

Prosper Tonderai Mataruse is a PhD candidate in the Department of Community and Social Development at the University of Zimbabwe. His research interests are in institutional aspects of natural resource management, climate change, social-ecological systems and their resilience.

Lindah Mhlanga is an environmental scientist (Biologist) with expertise and experience in freshwater and terrestrial ecosystems in Zimbabwe. Lindah has a strong research background and is experienced in issues of resource use governance and is well versed with governance and policy issues related to natural resources management in Zimbabwe. Through a book that she edited and co-authored (*Fragmentation of natural resources management: experiences from Lake Kariba.* PLAAS. Lit-Verlag Switzerland), she advocates for an ecosystem approach to natural resource management emphasizing the interrelatedness of the ecological and socio-economic dynamics of natural resources. She actively engaged in the outcome monitoring surveys under the Zimbabwe Resilience Building Fund project.

Michael Ticharwa Mubvuma is a senior lecturer and chairperson in the Department of Soil and Plant Sciences at Great Zimbabwe University (GZU). He chairs the GZU Climate Change Research Group. Michael holds a PhD from the University of Venda. His interests are in dryland agriculture and climate change.

Clarice Princess Mudzengi is an innovation hub manager at Great Zimbabwe University (GZU). She holds a PhD in Spatial Ecology and Remote Sensing. Her research interests are diverse, with the aim of achieving resilient and sustainable livelihoods in a changing climate.

Charity Mumaniki is a senior agricultural economist with the Ministry of Lands, Agriculture, Fisheries, Water and Rural Development. She is also a PhD candidate in the Department of Agricultural Business Development and Economics under the Faculty of Agriculture, Environment and Food Systems at the University of Zimbabwe. She holds a Master of Science Degree in Agricultural and Applied Economics and has more than 8 years' experience in agricultural policy planning, market and trade development and food value chains.

Andrew Mupwanyiwa is a lecturer in the Department of Economics at the University of Zimbabwe. He holds an MSc in Economics as well as a BSc in Economics. His research interests are in agricultural economics, environmental economics, health economics and development economics. He is working towards a PhD in Agriculture Economics.

Edward Mutandwa is an associate professor in the Department of Agricultural Economics and Extension at the University of Zimbabwe. He holds a PhD degree in Forest Economics and Policy from Mississippi State University, USA. He is also a holder of Master's and Bachelor's degrees in Agricultural Economics from the University of Zimbabwe. He has over 20 years' experience in the academic and development working spanning countries that include Zimbabwe, Rwanda and the United States. His research interests are in the areas of agricultural and environmental economics including ecosystem services valuation, urban forestry and food security-poverty analysis.

Emmanuel Mwakiwa is a senior lecturer and researcher in the Department of Agricultural Business Development and Economics in the Faculty of Agriculture, Environment and Food Systems at the University of Zimbabwe. Emmanuel has over 25 years' comprehensive experience in development planning, research and evaluation work. He has researched and written extensively in these areas in Africa with special emphasis in the southern Africa region and has published over 25 publications in high-impact referred international journals.

Teresa Nyika holds a PhD in Agricultural Economics from the University of KwaZulu-Natal, South Africa. She is a senior lecturer and the current chairperson of Economics and Development Department at the University of Zimbabwe. Her research experience and interests include public policy, gender and macroeconomics, economics of smallholder farming systems, household food security, adoption of agricultural technologies and climate-smart agriculture in agricultural development.

Kefasi Nyikahadzoi has a PhD in Applied Social Sciences from the University of Zimbabwe. He is the current director of the Environment, Climate and Sustainable Development Institute at the University of Zimbabwe. His research interests are in institutional economics, climate change, food security and innovation systems.

Carren Pindiriri is a senior lecturer/researcher in the Department of Economics at the University of Zimbabwe. He holds a PhD in Economics (UZ), MSc and BSc in Economics (UZ) and an MPhil in Environmental and Development Economics (UIO). He specializes in applied economics. His research interests are in development, agricultural economics, climate change and environmental economics. He has extensive knowledge in research both at national and regional level.

Xavier Poshiwa is a senior lecturer and dean of Gary Magadzire School of Agriculture at Great Zimbabwe University (Zimbabwe). He holds a PhD in Rangeland Ecology from Wageningen University (The Netherlands). His research interests are in biodiversity conservation, sustainable development, dryland agriculture and climate change.

Jayne Stack is an agricultural economist who has lived and worked in Zimbabwe for over 30 years. Formerly a senior lecturer in the Department of Agricultural Economics at the University of Zimbabwe, she currently provides Monitoring, Evaluation, Research and Learning support to resilience-related projects including the Zimbabwe Resilience Building Fund (ZRBF, 2018-2023), a multi-donor programme that sought to improve the resilience of about 830 000 people across 18 districts in Zimbabwe. Jayne was the Research and Resilience lead for ZRBF's Resilience Knowledge Hub and provided thought leadership on resilience approaches and anchored the team's technical thinking on multiple knowledge products and resilience tool kits. During that time, she also trained and supported other researchers, including university academics and graduate students, on understanding and analysing ZRBF quantitative and qualitative outcome monitoring data, to help surface evidence and build a repository of actionable learning and capacity building on what matter and works for resilience building.

Simbarashe Tatsvarei is a senior lecturer at the University of Zimbabwe since 2020. He holds a PhD in Agricultural Economics and an MBA, and has more than 20 publications and more than 20 years working experience. His research areas of interest are land economics, value chains and impact evaluations. Simbarashe is a reviewer of the following journals: *Land Use Policy*, *Agricultural Education and Extension*, *Agriculture and Food Security* and *GeoJournal*. He is a fellow of the African Science Leadership Program, and a member of the Social Science and Humanities Association.

Doreen Tsitsi Tirivanhu is an environment, climate change and resilience expert with 21 years' experience in successfully coordinating, conceptualizing and conducting interdisciplinary research on the environment and sustainable natural Resources management including research in urban and rural resilience. Tirivanhu holds a BSc in Agricultural Economics and an MSc in Environmental Policy and Planning from the University of Zimbabwe and is currently studying towards a PhD in Environmental Sciences at Rhodes University. She has served as the deputy director for the Centre for Postgraduate Studies and is currently the deputy registrar academic at the University of Zimbabwe.

Byron Zamasiya is a climate change and food security researcher. He has a PhD in climate change and adaptation, a Master's degree in Economics and a Bachelor of Science Honors degree from the University of Zimbabwe. Byron has many years of experience in research (action and participatory) and project implementation in the agricultural sector on research-led development projects in Southern Africa. His research interests include climate change, resilience building, adaptation, food security, integrated soil fertility management, agricultural markets and innovation systems.

Chapter 1
Introduction: Building Resilience to Climate Change in Selected Districts in Zimbabwe Experiences from the Zimbabwe Resilience Building Fund Programme

Lindah Mhlanga and Samuel Kusangaya

1.1 Introduction

1.1.1 Implications of Climate Change in Zimbabwe

The increased rates of global warming, consequently resulting in climate change, is now scientifically broadly recognised as a key global challenge. Narratives of climate change are now central not only to the development discourse but are also increasingly framing the understanding of key global challenges, such as food security, deforestation, desertification, health, population growth versus resource balance, high energy demand and poverty (Kusangaya et al., 2014). The African continent has been highlighted as particularly vulnerable to climate change due to its low adaptive capacity and vulnerability (Callaway, 2004). One of the most vulnerable regions in Africa is the southern African region, owing in part to lack of financial, institutional and technological capacity; low adaptive capacity; endemic poverty; low technology uptake and dependence on rain-fed agriculture (Kusangaya et al., 2014). However, despite Africa being one of the lowest contributors to greenhouse gas emissions, its developmental sectors are already experiencing widespread losses and damages attributable to climate change (IPCC, 2022).

The impacts of climate change have been noticeable in agriculture (Vermeulen et al., 2012), ecosystems and biodiversity and energy generation (Yamba et al., 2011). Patterns of weather-related hazards are affected by climate change and thus

---

L. Mhlanga (✉)
Environment, Climate and Sustainable Development Institute, University of Zimbabwe, Harare, Zimbabwe

S. Kusangaya
Department of Geography, Geospatial Sciences and Earth Observation, University of Zimbabwe, Harare, Zimbabwe

L. Mhlanga et al. (eds.), *Building Resilience to Climate Change in Selected Districts in Zimbabwe*, https://doi.org/10.1007/978-3-031-80529-5_1

have an impact on human settlements, causing loss of life, social disruption and economic hardships (Hulme et al., 2001). Such impacts will be felt more by the resource poor in Southern Africa (Ngigi & Denning, 2009). Negative impacts such as reduced food production, reduced economic growth, increased inequality and poverty, biodiversity loss, increased human morbidity, water shortages and loss of lives are projected to become widespread and severe with increased global warming (IPCC, 2022). Projected increases in the frequencies and magnitudes of such events through to 2050, coupled with increasing population in the region, suggest that the vulnerability of weather-sensitive sectors, such as agricultural production, will also increase (Yamba et al., 2011). IPCC (2022) reported that Zimbabwe is the second highest climate disaster risk country in Africa after Mozambique due to the increasing frequencies of droughts and floods.

Tropical cyclone activity in Southern Africa has been recorded with increased frequency in recent years, causing massive damage and loss of life mainly due to flooding (Reason & Keibel, 2004). Recurring floods (Artur & Hilhorst, 2012) have resulted in food insecurity (Tirado et al., 2010) and exacerbated health conditions, loss of shelter and even loss of life as happened during the 2000–2001 Cyclone Eline, and the 2019 cyclone Idai induced floods, which affected Mozambique, Zimbabwe and South Africa (Chanza et al., 2020). Tropical Cyclone Idai hit Manicaland Province in Zimbabwe on 15–17 March 2019, resulting in disrupted livelihoods, destruction of houses, infrastructure damage and loss of agricultural land (Mazvimavi, 2019).

Climate change, increasing climate variability and the consequent impact are already being experienced in Zimbabwe as evidenced by increasing temperatures and erratic rainfall and changing rainfall patterns (World Bank, 2020). Clear trends of increasing temperatures over Zimbabwe since 1900 have been established with temperatures having risen by 0.4 °C in Zimbabwe (Brazier, 2019). Temperature increases are now apparent for the dry seasons with seasonal warming now being experienced in Zimbabwe.

Rainfall patterns have significantly changed over the last few decades, with a slight decrease in total rainfall noted (Girvetz et al., 2019) between 5% and 18% with increasing variability and frequent incidences of droughts, floods and storms (Brazier, 2019). Increased frequency of mid-season dry spells and altered patterns of precipitation are now a characteristic feature in Zimbabwe (Mupangwa et al., 2011), which has adversely impacted the livelihoods of approximately 70% of the country's rural population that is dependent on rain-fed agriculture (Muzari et al., 2016). Several extreme weather events are now frequently being experienced, ranging from droughts and dry spells, tropical cyclones and floods, to ground frost and heatwaves.

The impact of climate change is very apparent for small holder agricultural production due to the strong coupling between the success of the country's agricultural system and rainfall, and the ensuing vulnerability of this sector to dry spells and droughts (UNDP, 2020). The projected general decrease in precipitation will have a significant impact on livelihoods. Since most livelihoods are dependent on rain-fed agriculture, they will be affected by a reduction in rainfall and an increase in temperatures under climate change (Thornton et al., 2010). Rainfall is projected to be more variable and erratic and the rainy season onset dates are projected to change and become more irregular, whilst dry periods will become more frequent and longer (Red Cross Climate Centre, 2021). Recurrent drought will be prevalent, especially in the southwestern provinces (Frischen et al., 2020). Smallholder agricultural production will be constrained by high temperatures and low rainfall (Mano & Nhemachena, 2007), thereby compounding peasant farmers' climate vulnerability to drought, which would result in food insecurity and poverty.

The revision of Zimbabwe's Agroecological Zones (AEZs) by Manatsa et al. (2020) showed that due to climate change the previous AEZs were no longer in tandem with the expected agricultural productivity. Climate change has resulted in changes in temperature and rainfall regimes, thereby affecting agricultural practice relevant within different agro-ecological regions of the country (Mugandani et al., 2012; Manatsa et al., 2020). According to Manatsa et al. (2020), there have been significant spatial shifts in AEZ, shifting mostly towards drier and less productive categories. As a result, new agroecological zones, such as natural region Vb, were added and occupy approximately 9% of the country. Agroecological region Vb is regarded as a region not suitable for sustaining any form of rain-fed agriculture including growing of drought-tolerant crops. According to Manatsa et al. (2020), the classification into natural regions Va and Vb was reflective of the increased aridity, mostly driven by climate change, specifically, the significant reduction in rainfall coupled with increased dry spells and high evapotranspiration, which all reduces effective moisture, rendering the region too dry for viable agriculture.

1.1.2 Importance of Water Under a Changing Climate and Implications on Livelihoods

Zimbabwe is characterised by highly spatial and temporally variable rainfall and, in some cases, scarce water resources (Kusangaya et al., 2014). A change in climate could affect water supplies and, hence, livelihoods. Water is the primary medium through which climate change impacts are felt (Caretta et al., 2022; UNESCO-UN-Water, 2020). Shortage of water has negative effects on the livelihoods of the people in low-income rural communities of Zimbabwe (ZRBF-ECRIMS, 2019), which are largely water dependent.

Emissions scenarios project decreased availability of renewable surface and groundwater resource in Southern Africa. Water will continue to be scarce because each degree of global warming is projected to decrease the availability of renewable water resources by at least 20%. Climate change has the potential to worsen water availability in Zimbabwe. It has been projected that by the year 2050, Zimbabwe will be in the high-water stress category even without climate change (Mazvimavi, 2010). As such, climate change is expected to exacerbate current stresses on water resources, impacting availability and access to water resources and the sustainability of local ecosystems (Kusangaya et al., 2021).

Socio-economic drivers such as population growth, economic development and associated increases in per capita water use and expansion of agricultural and industrial activities exacerbate the risk of water scarcity (IPCC, 2014). Other economic sectors such as energy generation, tourism, health, ecosystems and biodiversity will have to adapt to scarce water resources (Kusangaya et al., 2021). Water adaptation technologies that can alleviate climate change–related water problems are already being adopted in dry and semi-arid parts of Zimbabwe. Development and implementation of technologies for adapting to water scarcity, such as rainwater harvesting and conservation agriculture, is central not only to sustainable development but also to the survival of most communities in sub-Saharan Africa. Adaptation in the water sector could be done through improved water conservation, enhanced efficiency, technological change and adoption of integrated water resources management (Schulze, 2010).

Livelihood activities among rural communities are centred on rain-fed agricultural production and, as such, communities are highly susceptible to climate-related hazards and shocks (ZimVAC, 2016). Climatic changes, especially in the drier parts of Zimbabwe, will aggravate existing food insecurities in communal areas by affecting agricultural productivity (Chikodzi et al., 2012). Food insecurity, water scarcity and loss of livestock will increase the vulnerabilities of communities. It is now difficult to accurately depict the onset of the planting season due to highly unpredictable, erratic and unevenly distributed rainfall patterns, while the prevalence of mid-season dry spells and frequent below normal rainfall are causing crop failures and, thereby, exacerbating food insecurity (Mutekwa, 2009). Increases in drought frequency and duration are projected over large parts of Southern Africa above 1.5 °C global warming (IPCC, 2022). Thus, water availability will become a challenge, especially for agriculture production, a primary source of livelihood for 70% of Zimbabwe's population (ZimVAC, 2018).

Besides climate change, rain-fed agricultural production, purchasing power and household food security are also adversely affected by environmental, economic and social shocks and stresses (Sam et al., 2019). Climatic induced environmental shocks aggravate these, thus increasing the vulnerability of rural households and communities in Zimbabwe (Mutimukuru-Maravanyika et al., 2022). In addition, shocks such as global recession, hyperinflation, diseases and market instability also increase the vulnerability of the rural households and communities in Zimbabwe (Harvey et al., 2018) as they become less able to recover from the climate shocks, such as floods and droughts (Kom et al., 2020).

1.1.3 Nature of Shocks Experienced in Zimbabwe

Ngenoh et al. (2018) categorised shocks into (1) production shocks and (2) marketing shocks. Production shocks include weather-related (droughts, storms, floods and landslides), natural disasters (flood, storm, unusually heavy rain, landslides/ erosion) and biological (pests and diseases). On the other hand, marketing shocks are categorised into changes in market demand and supply, instabilities, fuel price hikes and general price volatilities (Ngenoh et al., 2018). These shocks, in total, lower prices of produce and result in loss of income and production of low-quality products (Mabuza et al., 2016). Food markets and local economies are affected by these shocks, thus affecting food security. The most severe shocks experienced in Zimbabwe are drought, mid-season dry spells, flooding, human diseases, prices of cereals and livestock, crop pests and diseases, animal diseases, economic shocks such as price inflation on food and farming inputs and loss of employment (UNDP, 2021). In Zimbabwe the effect of tropical cyclones, resulting in floods, landslides and diseases (cholera and malaria) is most profound. Recurring floods have resulted in food insecurity, exacerbated health conditions, loss of shelter and even loss of life (Chanza et al., 2020). Cyclone Eline caused 91 deaths, 357 injuries, destroyed 59,187 houses and resulted in about 2.7 million people being affected (Chanza et al., 2020). Cyclone Idai changed the lives and livelihoods of an estimated 270,000 people in Chimanimani District and caused between 299 and 344 deaths, while more than 300 people were reported missing (Mazvimavi, 2019).

1.1.4 What and Why of Resilience?

Building resilience is about enhancing and reinforcing the capacities, livelihoods and opportunities of the most vulnerable and food-insecure people, communities and countries in the face of an increasingly risky environment (Mwadzingeni et al., 2022). Resilience is defined from a social-ecological perspective as the capacity of socio-economic systems (e.g., households) to withstand shocks through absorption, adaptation and transformation. Anti-poverty programmes have prioritised building household resilience to cushion households exposed to new or increasingly severe climate and economic shocks. Resilience capacity is an expression of the household's absorptive, adaptive and transformative domains (Kebede et al., 2016). Resilience can be understood from the perspectives that include: the "capacity to resist shock and bounce back" and ecological resilience (adaptive capacity); the "capacity to buffer, adjust and continue functioning following shocks" to include the (transformative capacity); the "capacity to create a fundamentally new system in times of crises" (Asmamaw et al., 2019). Resilience is a broad-based process that can be observed over temporal and spatial scales (analysed in this book over a three-year period).

The definition of resilience is "*the ability of at-risk individuals, households, communities and systems to anticipate, cushion, adapt, bounce back better and move on from the effects of shocks and hazards in a manner that protects livelihoods and recovery gains, and supports sustainable transformation*" (UNDRR, 2015). The resilience approach focuses on how natural resources can be managed efficiently and sustainably in the face of disturbances and uncertainty. The following five resilience principles were agreed in the Resilience Strategic Framework for Zimbabwe: (i) focus on shock dynamics, (ii) resilience as a multidimensional capacity, (iii) resilience functions, (iv) outcome-indexed capacities and (v) multilevel and systems-based. The ZRBF programme sought to ensure that development programmes are designed and based on evidence concerning the shocks and stresses that affect the well-being of rural communities in Zimbabwe, as well as their capacities to withstand and recover from these shocks (www.zrbf.co.zw/about/approach).

1.1.5 The Role of Livelihood Assets, Especially Livestock Assets in Resilience Building

In building household resilience, assets are crucial in facilitating response to hazards. Different assets can improve resilience before, during and after an adverse event. It is consequently critical to retool smallholder farmers with productive assets when building resilience (Chigunhah et al., 2020). Assets increase farmers' ability to bounce back to their original state or, better still, help them become better off than they were before (Hendriks, 2015). Livestock assets play a central role on smallholder farmers' resilience to shocks (Ngigi et al., 2015) by acting as a source of wealth and a means to sustain livelihoods in the face of environmental and socio-economic shocks. Ownership of productive assets such as livestock plays an important role in resilience building by acting as financial reserve buffers for smoothening consumption during a shock (Nelson et al., 2016). Goats and cattle can be disposed of during shocks and stresses to supplement household food needs (Zamasiya et al., 2021) thereby offering an opportunity to withstand a shock or stress (Nelson et al., 2016). Livestock are, however, also threatened by climate shocks especially droughts, which impact the quality and quantity of forage and water availability, thereby making them prone to diseases and poverty deaths. Livelihood diversification assists households in dealing with the effects of shocks and stresses on a household's resilience (Nelson et al., 2016). By venturing into non-farm or farm-based livelihood strategies, households can become less vulnerable.

1.1.6 Vulnerability

Differential vulnerability analysis in ZRBF programming highlights some of the primary gender and age-related limitations that exclude certain groups from full access and use of resilience capacities, an important element to consider when designing resilience pathways. The ZRBF Youth Integration Strategy proposes the following areas of change that are likely to reverse these limitations and improve the inclusion of marginalised and vulnerable groups in resilience pathways: (i) increased proportion of resilience assessments that evaluate different vulnerabilities of men, women and youth and as a basis for interventions; (ii) increased leadership and meaningful participation of women and youth in resilience-building structures and activities; (iii) improved access, control and decision-making over information, services, markets and outputs related to women's and youth resilience; and (iv) fairer division of labour between women, men and young people in balancing household responsibilities and building resilience.

Climate vulnerability, the degree to which a system is susceptible to, and unable to cope with, adverse effects of climate change and variability and extremes (IPCC, 2007), is a function of climate sensitivity and adaptive capacity of communities and natural systems (Xia et al., 2017). In rural Africa, poor and female-headed households face greater livelihood risks from climate hazards (IPCC, 2022). The impacts of climate change on health, livelihoods and food security of different groups are felt in different ways. Nevertheless, vulnerability is anticipated to worsen for marginalised communities who are already struggling to manage other shocks and stressors. If the greenhouse gas emissions continue to rise unabated under a business-as-usual scenario, the average temperature for Africa will increase by 2.6–4.8 °C by the end of the twenty-first century (IPCC, 2014). Droughts, economic deterioration and social shocks increase the vulnerability of marginalised households (Koomson et al., 2020) by limiting abilities to generate income and access markets, eroding savings and forcing the disposal of productive assets to finance food security requirements. As vulnerability increases the capacity to respond to future shocks and stresses is hindered and food security is affected.

It is necessary to invest in young people because they are future leaders and workforce (Berliner, 2005). In Zimbabwe, young people are often excluded from rural agriculture because they lack access to productive resources such as land, markets and finance (Mudege & Kwangwari, 2013). However, young people are also negatively affected by such shocks because they cause loss of assets on which they depend, leading to a vicious cycle of poverty and vulnerability, which they may fail to break unless appropriate interventions are instituted. Unless programming is properly informed, existing gender or age-related inequalities can be reinforced by shocks and stresses. Unless addressed, drought and extreme weather impact will increase the burden on female smallholder farmers dependent on rain-fed agriculture and climate-sensitive resources (Brown et al., 2012). Women and youths participate in non-farm resilience-building initiatives (UNDP, 2021). Youth could be engaging in non-farm activities as a livelihood source in response to the effect of different shocks and stresses on agricultural production (Nandi et al., 2022). There has to be an understanding of the challenges encountered by youths and women in participating in resilience activities which include insufficient access to knowledge, information and education, limited access to land, inadequate access to financial services, difficulties accessing green jobs, limited access to markets and limited involvement in policy dialogue (FAO, 2014). Differentials noted include the increase in the participation of female youths participating in savings groups (UNDP, 2021), while male youths preferred to engage in gold mining as a coping strategy in times of shocks and stresses, whereas others opted for skilled trades and artisanal livelihood sources, such as carpentry, welding, sewing and detergence making (UNDP, 2020).

1.1.7 Climate Smart Agriculture

Climate change impacts food security, and in Zimbabwe, food insecurity poses a major threat to 7.7 million people due to poor rains and erratic weather patterns, which have negatively impacted crop harvests and livelihood prospects (WFP, 2020). Climate Smart Agriculture (CAS) pathways are being adopted in order to build up higher resilience and lower risks to food insecurity. CSA practises can improve household food security under climate variability by increasing agricultural income (through higher yields or lower production costs), thus increasing the resilience of households to absorb shocks (Wainaina et al., 2018). There is a need for inclusion of a gender-responsive approach to programming so as to enhance the three main pillars of CAS, namely, (i) sustainably increasing agricultural productivity and incomes; (ii) adapting and building resilience to climate change; and (iii) reducing and/or removing greenhouse gas emissions, where possible, in order to improve food and nutrition security (Nelson & Huyer, 2016).

1.1.8 Zimbabwe Resilience Building Fund

UNDP, DFID, EU and Sweden funded the setting up of the Zimbabwe Resilience Building Fund (ZRBF) in 2015. The Ministry of Lands, Agriculture, Water, Climate and Rural Resettlement (MLAWCRR) implemented ZRBF for the Government of Zimbabwe. ZRBF was aimed at contributing to the increased capacity of at-risk communities to protect development gains and achieve improved well-being in the face of shocks and stresses (UNDP, 2020). A core focus of ZRBF was to build the resilience of individuals, households, communities and systems through the following interconnected components: (a) building evidence; (b) supporting long-term household and community resilience; and (c) a crisis modifier mechanism that can provide agile and flexible funding in times of shocks and stresses.

The ZRBF was founded to address economic, environmental and social shocks and stresses affecting rural communities in Zimbabwe. The ZRBF theory of change was based on the hypothesis that if investments are made to: (i) directly support targeted communities to improve absorptive, adaptive and transformative capacities; (ii) avail timely and cost-effective response to emergencies rolled out; and (iii) generate learning and utilise evidence in policies and decisions; then not only will targeted communities be more resilient and food secure, but there will also be a better understanding of what works and what does not work to strengthen communities' resilience in Zimbabwe (UNDP, 2020). It is anticipated that the ZRBF project would assist by increasing resilience capacities, improving well-being outcomes and aid in recovering from climate related shocks.

The ZRBF was a coalition of donors supporting resilience-building activities of organisations operating in 7 consortia within 18 districts of Zimbabwe (Fig. 1.1, Table 1.1). The ZRBF defined resilience as the ability of at-risk individuals, households, communities and systems to anticipate, cushion, adapt, bounce back better and move on from the effects of shocks and hazards in a manner that protects livelihoods and recovery gains and promotes long-term transformation. The main principles under ZRBF were that (i) investments should directly support targeted communities so that they would improve their resilience capacities (absorptive, adaptive and transformative); (ii) timely and cost-effective emergency response is provided; and (iii) evidence is used to make decisions and inform policies, then targeted communities will be more resilient and food secure, and knowledge

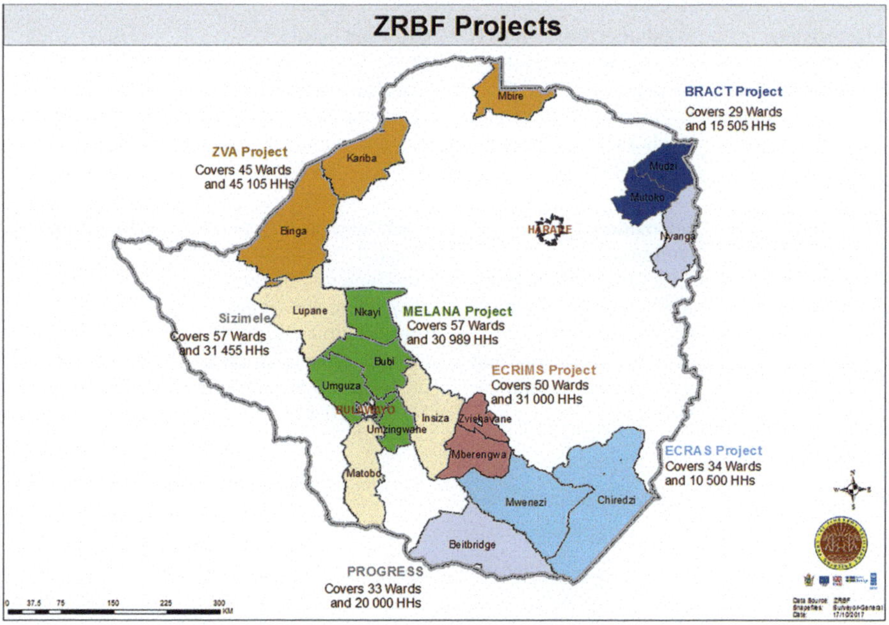

Fig. 1.1 ZRBF Consortia and districts. (Source: zrbf.co.zw/projects)

Table 1.1 The seven consortia supporting the ZRBF Programme. (Source: zrbf.co.zw/projects)

Consortium		Initiatives	Districts
Building Resilience through improving the Absorptive Capacity for Transformation (BRACT)	Christian aid Silveira House (SH), Bio-Innovation Zimbabwe (BIZ), Community Technology Organisation (CTDO), Nyahurune Community Trust (NCT)	Livestock production, Crop production and apiculture, Non-timber and Forest Products, Early Warning System, Private Sector engagement/market linkages, Climate adaptation, strengthening existing social savings groups and new ISAL groups and SACCOs, Linking rural SMEs	Mutoko Mudzi 29 wards 15,505 households
Enhancing Community Resilience and Sustainability (ECRAS)	CARE PLAN ICRISAT (international crops research institutes for semi-arid tropics)	Crop and livestock production, productive assets, early warning systems, private sector engagement/market linkages, climate adaptation	Chiredzi, Mwenezi 34 wards 10,500 households
Enhancing Community Resilience and Inclusive Market Systems (ECRIMS)	CARE Lutheran Development Services (LDS), International Crops Research Institute for the Semi-Arid Tropics (ICRISAT), Local Initiatives for Development Agency (LID)	Crop and livestock production, apiculture, private sector/market linkages, climate adaptation, early warning systems, Village Savings and Lending Association (VSLA), commodity groups, linking VSLAs	Mberengwa, Zvishavane 50 wards 31,000 households
Matabeleland Enhanced Livelihoods, Agriculture and Nutrition Adaptation (MELANA)	Welthungerhilfe (WHH) Community Technology Organisation (CTDO), Institute of Environmental Studies (IES), Agricultural Partnerships Trust	Crop and livestock production, apiculture, early warning system, private sector engagement/market linkages, climate adaptation, productive assets	Nkayi, Bubi, Umguza, Umzingwane 54 wards, 30,989 households
Programme for Growth and Resilience (PROGRESS)	International rescue committee CESVI, BIOHUB trust, MRI (Matopos research institute), CIMMYT (international maize and wheat improvement Centre)	High value cash cropping, crops, livestock, private sector engagements, climate adaptation, non-timber forestry products, mechanisation and service provision, ISAL linkages with MFIs and banks	Beitbridge and Nyanga 33 wards 20,000 households

(continued)

Table 1.1 (continued)

Consortium		Initiatives	Districts
Sizimele action for building resilience in Zimbabwe (ZIZIMELE)	DanChurchAid (DCA) Institute for Rural Technologies (IRT) Organisation of rural technologies, Techno serve, community based organisations, ProAfrica community based organisations, CyVAT, Community Capacity, Building Initiative for CCBICA, Future of Hope Foundation (FoHF), Midlands State University (MSU), HWA	Crop and livestock production, district level processing centres, small livestock/cattle finishing posts, private sector engagements, climate adaptation, insurance and productive assets early warning systems, ISALS, linkages with MFIs, banks, impact investors, cattle banking, risk insurance	Matobo, Insiza, Lupane 57 wards 31,455 households
Zambezi Valley Alliance for building community (ZVA)	Action aid Zimbabwe Environmental Law Association (ZELA), African Breeders Services Total Cattle Management (ABS TCM)	Crop and livestock production, entrepreneurship and apiculture, WASH, early warning system, private sector engagements/ market linkages, Productive assets, Climate adaptation	Binga, Mbire, Kariba 45 wards 45,105 households

generated will lead to a better understanding of how to build community and household resilience.

The concept of resilience provides a conceptual framework that is helping to understand how communities can adapt to climate change (UNDP, 2021). Resilience building strategies can counter vulnerability to shocks (UNDP, 2021). The severity and magnitude of livelihood shocks and stressors due to changing climate can be minimised through resilience building among communities. Most development projects are now incorporating resilience building (Frankernberger et al., 2013) by investing in adaptive, absorptive and transformative capacities of households and communities (Bene et al., 2012). Building the resilience of households and communities living in arid and semi-arid regions where the impact of climate change will be high will avert challenges in the face of the predictions of worsening future climate change (IPCC, 2014). Development actors can lessen the effects of shocks and stressors by embedding resilience in their interventions. Early and proactive adaptation and resilience building actions are more cost-effective than addressing impacts after they occur.

The logic behind ZRBF resilience programming was that programming influences practices and behaviours, which in turn change resilience capacities, leading to improved food security and well-being outcomes in the face of shocks. One of the ways that programming influences resilience capacities is through improving the effectiveness of livelihood assets (training, exposure to information) and supporting the accumulation of assets by individual households and communities (including social capital through the formation of farmer groups). The three critical capacities (absorptive capacity, adaptative capacity and transformative capacity) are interconnected, mutually reinforcing and exist at multiple levels (individual, household, community, national and ecosystem levels).

ZRBF interventions were aimed at improving the resilience of households in 18 districts in Zimbabwe. Livelihood diversification through resilience building interventions enables communities to adapt to climate changes by enhancing household and community absorptive, adaptive and transformative capacities. The primary resilience pathways of ZRBF include a variety of interventions aimed at developing absorptive, adaptive and transformative resilience capacities capable of effectively preventing and mitigating shocks and stresses. The ZRBF programme focused on building household and community resilience and focused on a population of about 800,000, predominantly located in the country's arid and semi-arid regions. The programme was anchored on three interrelated components namely;

- Component 1: Gathering data to strengthen the regulatory framework and encourage the delivery of services which will increase the resilience of homes and communities.
- Component 2: Long-term household and community resilience is supported by interventions that emphasise developing capacities for absorption, adaptation and transformation.
- Component 3: A risk financing mechanism that provides target communities experiencing humanitarian shocks with predictable, adequate and timely funding.

Each of the aforementioned goals works to increase households' resilience to unforeseen climate change events and natural stresses. The seven consortia supported different initiatives or practices, but they all focused on building household social capital (Fig. 1.1, Table 1.1).

1.1.9 The Purpose of the Book

The book is based on the ZRBF data from the Outcome Monitoring Surveys 1,2,3 (OMS123), which comprised three rounds of panel data set with the same households being interviewed three consecutive times in 18 districts (Fig. 1.1). The OMS1, OMS2 and OMS3 were undertaken in 2019, 2020 and 2021, respectively. It is anticipated that the information contributed in each chapter will be useful in informing adaptive programming and design of future resilience interventions and policy. The findings in this book demonstrate and provide a better understanding of what works and what does not work to strengthen communities' resilience in Zimbabwe (UNDP, 2020). The findings can be shared with the Climate Resilient Development (CRD) in Zimbabwe. CRD is a process of implementing GHG mitigation and adaptation measures to support sustainable development for all. It addresses the challenge of paucity in climate-related research in programming in Zimbabwe (severe data constraints).

1.1.10 Outline and Organisation of the Book

The book has 8 chapters. In Chap. 2, Nyika et al. examines the impact of climate smart agriculture on gender differentiated food and income security outcomes in Zimbabwe. Chuma et al. used a mixed approach research design, which combines the use of quantitative and qualitative approaches to evaluate the determinants of Climate Smart Agriculture (CSA) adoption among women and men and measures and evaluates the impact of CSA practices and ZRBF interventions on beneficiaries' incomes and food security status. The impacts of CSA practices on food and income security in Zimbabwe were evaluated with a gender lens.

In Chap. 3, Zamasiya et al. assess the determinants of household food security among smallholder farmers under changing climates and contexts using a case of households in the Zambezi Valley in Zimbabwe. The chapter focuses on how adaptive capacity drives household food security. Building adaptative capacity strengthens households' ability to make informed and proactive decisions about their livelihood choices such as preparing for climate-related risks to their productive activities (UNDP, 2021). The Ordinary Least Squares (OLS) regression was used to investigate the drivers of household food security for smallholder farmers facing climate-induced shocks and stresses.

In Chap. 4, Bhatasara and Karuma describe water-based infrastructure and how the different water related resilience capacities are contributing to building resilience outcomes. The authors used a thematic analysis approach. With multiple layers and an emphasis on induction, thematic data analysis allowed themes, issues, trends, patterns and conclusions to emerge from the data. Based on the idea of resilience capacities—resources that an individual, family, community, or system can access and employ in a particular way to deal with shocks and stresses in order to lessen or eliminate their effects on developmental progress—the analysis in this chapter is conducted. In this case, the chapter examines the results of employing resilience capacities as it addresses pathways to resilience using water-based technology.

In Chap. 5, Makiwa et al. investigate the contributions of ZRBF in the building of rural household resilience, focusing on how livelihood assets are used to generate sustainable resilience capacity flows for the household. The sustainable livelihoods framework was used by applying the random effects panel probit model to assess the relationship between resilience capacity and the underlying socio-economic and demographic predictors. The sustainable livelihoods framework permits the examination of complex relationships between resilience, household assets and critical demographic. Makiwa et al. adopted the ZRBF resilience indexing methodology, where household resilience is an amalgamation of absorptive, adaptive and transformative capacities based on numerous socio-economic and demographic variables.

In Chap. 6, Mudzengi et al. examine the significance of livestock-based assets in building resilience for 3440 rural households across dryland agro-ecological zones in Zimbabwe. The main objective of this study was to establish the role of livestock in building resilience against four shocks, namely drought, low livestock prices, livestock deaths due to diseases and livestock theft. In the context of a changing climate in Zimbabwe, it is critical that the role of livestock in building resilience against environmental and socio-economic shocks is clearly defined.

In Chap. 7, Manjengwa et al. used the experience of beneficiary households of the ZRBF programme to better understand the role of resilience capacity building as a plausible pathway for reducing vulnerability of resource-poor rural farmers and the characteristics of beneficiary households who are achieving sustained escapes out of poverty. They focused on determining the relationship between social-economic status (extremely poor, poor and non-poor) and resilience capacities of rural households involved in the ZRBF Programme, the characteristics and factors of the beneficiary households which have both increased resilience capacity and improved socio-economic status and the resilience pathways/resilience interventions of the sustained poverty escapers.

In the concluding chapter, Chap. 8, Zamasiya & Nyikahadzoi synthesise key findings and propose policy interventions to bolster smallholder farmers' resilience against climate change and other extreme events. This chapter summarises the main findings from the case studies and proffers recommendations for enhancing the resilience of smallholder farmers to social, economic and environmental shocks and stresses.

References

Artur, L., & Hilhorst, D. (2012). Everyday realities of climate change adaptation in Mozambique. *Global Environmental Change, 22*(2), 529–536.

Asmamaw, M., Mereta, S. T., & Ambelu, A. (2019). Exploring households' resilience to climate change-induced shocks using climate resilience index in Dinki watershed, central highlands of Ethiopia. *PLoS One, 14*(7), e0219393. https://doi.org/10.1371/journal.pone.0219393

Bene, C., Wood, G. R., Newsham, A., & Davies, M. (2012). *Resilience: New utopia or new tyranny? Reflection about the potentials and limits of the concept of resilience in relation to vulnerability reduction programmes*. Institute of Development Studies.

Berliner, D. (2005). An impossible transmission: Youth religious memories in Guinea-Conarky. *American Ethnologist, 32*(4), 576–592.

Brazier, A. (2019). Climate change in Zimbabwe. In *A guide for planners and decision-makers*. Konrad-AdenauerStiftung.

Brown, C., Ghile, Y., Laverty, M., & Li, K. (2012). Decision scaling: Linking bottom-up vulnerability analysis with climate projections in the water sector. *Water Resources Research, 48*, W09537. https://doi.org/10.1029/2011WR011212

Callaway, J. M. (2004). The benefits and costs of adapting to climate variability and climate change. *The Benefits of Climate Change Policies*, 111–157.

Caretta, M. A., Mukherji, A., Arfanuzzaman, M., Betts, R. A., Gelfan, A., Hirabayashi, Y., Lissner, T. K., Liu, J., Lopez, G. E., Morgan, R., Mwanga, S., & Supratid, S. (2022). Water. In *Climate change 2022: Impacts, adaptation, and vulnerability. Contribution of working group II to the sixth assessment report of the IPCC*. Cambridge University Press.

Chanza, N., Siyongwana, P. Q., Williams-Bruinders, L., Gundu-Jakarasi, V., Mudavanhu, C., Sithole, V. B., & Manyani, A. (2020). Closing the gaps in disaster management and response: Drawing on local experiences with cyclone Idai in Chimanimani, Zimbabwe. *International Journal of Disaster Risk Science, 11*(5), 655–666.

Chigunhah, B. R., Svotwa, E., Govere, I., & Chikazhe, L. (2020). Stimulating farmer access to bank credit in Zimbabwe: The bankers' perspective. *Journal of Economics and International Finance, 12*(3), 84–94.

Chikodzi, D., Simba, F., & Murwendo, T. (2012). Climate change scenarios, perceptions and crop production: A case study of semi-arid Masvingo Province in Zimbabwe. *Journal of Earth Science Climate Change., 33*. https://doi.org/10.4172/2157-7617.1000124

FAO. (2014). *Youth and agriculture: Key challenges and concrete solution*. FAO.

Frankenberg, E., Sikoki, B., Sumantri, C., Suriastini, W., & Thomas, D. (2013). Education, vulnerability and resilience after a natural disaster. *Ecology and Society, 18*(2), 16. https://doi.org/10.5751/ES-05377-180216

Frischen, J., Meza, I., Rupp, D., Wietler, K., & Hagenlocher, M. (2020). Drought risk to agricultural Systems in Zimbabwe: A spatial analysis of Hazard, exposure, and vulnerability. *Sustainability, 12*(3), 752. https://pdfs.semanticscholar.org/00df/5d00846c655355579170e86f490dcc4b88c3.pdf

Girvetz, E., Ramirez-Villegas, J., Claessens, L., Lamanna, C., Navarro-Racines, C., Nowak, A., & Rosenstock, T. S. (2019). Future climate projections in Africa: Where are we headed? In *The climate-smart agriculture papers* (pp. 15–27). Springer.

Harvey, C. A., Saborio-Rodríguez, M., Martinez-Rodríguez, M. R., Viguera, B., Chain-Guadarrama, A., Vignola, R., & Alpizar, F. (2018). Climate change impacts and adaptation among smallholder farmers in Central America. *Agriculture and Food Security, 7*, 57. (2018). https://doi.org/10.1186/s40066-018-0209-x

Hendriks, S. L. (2015). The food security continuum: A novel tool for understanding food insecurity as a range of experiences. *Food Security, 7*(3), 609–619.

Hulme, M., Doherty, R., Ngara, T., New, M., & Lister, D. (2001). African climate change: 1900-2100. *Climate Research, 17*(2), 145–168.

IPCC. (2007). *Climate change 2007: Impacts, adaptation and vulnerability. Contribution of working group II to the fourth assessment report of the intergovernmental panel on climate change*. Cambridge University.

IPCC. (2014). Freshwater resources. In B. E. J. Cisneros, T. Oki, N. W. Arnell, G. Benito, J. G. Cogley, P. Döll, T. Jiang, & S. S. Mwakalila (Eds.), *Climate change 2014: Impacts, adaptation, and vulnerability. Part A: Global and sectoral aspects. Contribution of working group II to the fifth assessment report of the intergovernmental panel on climate change* (pp. 229–269). Cambridge University Press.

IPCC. (2022). In H.-O. Pörtner, D. C. Roberts, M. Tignor, E. S. Poloczanska, K. Mintenbeck, A. Alegría, M. Craig, S. Langsdorf, S. Löschke, V. Möller, A. Okem, & B. Rama (Eds.), *Climate Change 2022: Impacts, Adaptation and Vulnerability. Contribution of Working Group II to the Sixth Assessment Report of the Intergovernmental Panel on Climate Change* (pp. 1285–1455). Cambridge University Press. https://doi.org/10.1017/9781009325844.011

Kebede, T., Haji, J., Legesse, B., & Mammo, G. (2016). Econometric analysis of rural households' resilience to food insecurity in west Shoa. *Ethiopia. Journal of Food Security, 4*(3), 58–67. https://doi.org/10.12691/jfs-4-3-2

Kom, Z., Nethengwe, N. S., Mpandeli, N. S., & Chikoore, H. (2020). Determinants of small-scale farmers' choice and adaptive strategies in response to climatic shocks in Vhembe District, South Africa. *GeoJournal, 87*, 677–700. (2022).

Koomson, I., Villano, R. A., & Hadley, D. (2020). Effect of financial inclusion on poverty and vulnerability to poverty: Evidence using a multi-dimensional measure of financial inclusion. In *Network for socio-economic research and advancement working paper, NESRA/WP/20/001* (pp. 1–25).

Kusangaya, S., Warburton, M. L., Van Garderen, E. A., & Jewitt, G. P. (2014). Impacts of climate change on water resources in southern Africa: A review. *Physics and Chemistry of the Earth, Parts a/b/c, 67*, 47–54.

Kusangaya, S., Mazvimavi, D., Shekede, M. D., Masunga, B., Kunedzimwe, F., & Manatsa, D. (2021). Climate change impact on hydrological regimes and extreme events in southern Africa. In S. Diop, P. Scheren, & A. Niang (Eds.), *Climate change and water resources in Africa*. Springer. https://doi.org/10.1007/978-3-030-61225-2_5

Mabuza, M. L., Ortmann, G. F., Wale, E., & Mutenje, M. J. (2016). The effect of major income sources on rural household food (in) security: Evidence from Swaziland and implications for policy. *Ecology of Food and Nutrition, 55*, 209–230.

Manatsa, D., Mushore, T. D., Gwitira, I., Wuta, M., Chemura, A., Shekede, M. D., Mugandani, R., Sakala, L. C., Ali, L. H., Masukwedza, G. I., Mupuro, J. M., & Muzira., N.M. (2020). *Revision of Zimbabwe's agro-ecological zones*. Government of Zimbabwe.

Mano, R., & Nhemachena, C. (2007). Assessment of the economic impacts of climate change on agriculture in Zimbabwe: A Ricardian approach. In *Policy research working paper 4292*. World Bank.

Mazvimavi, D. (2010). Climate change, water availability and supply. *SARUA Leadership Dialogue Series, 2*(4), 81–97.

Mazvimavi, D. (2019). Cyclones Idai and Kenneth in South African Region in 2019. In K. Hiroki (Ed.), *HELP global report on water and disasters 2019, Secretariat of high-level experts and leaders Panel on water and disasters (HELP).* https://www.wateranddisaster.org/cms310261/wp-content/uploads/2019/07/HELP-Global-Report-on-Water-and-Disasters-D9-20190607_s.pdf

Mudege, N. N., & Kwangwari, C. (2013). Women and participation in civil society: Do women get empowered? The case of women in Goromonzi District in Zimbabwe. *Journal of Women, Politics and Policy, 34*(3), 238–260. https://doi.org/10.1080/1554477X.2012.722429

Mugandani, R., Wuta, M., Makarau, A., & Chipindu, B. (2012). Re-classification of agro-ecological regions of Zimbabwe in conformity with climate variability and change. *African Crop Science Journal, 20*, 361–369.

Mupangwa, W., Walker, S., & Twomlow, S. (2011). Start, end and dry spells of the growing season in semi-arid southern Zimbabwe. *Journal of Arid Environments, 75*(11), 1097–1104.

Mutekwa, T. (2009). Climate change impacts and adaptation in the agricultural sector: The case of smallholder farmers in Zimbabwe. *Journal of Sustainable Development in Africa, 11*(2), 237–256.

Mutimukuru-Maravanyika, T., Chapungu, L., & Majeke, F. (2022). Food Systems in Dryland Communities: Challenges and opportunities in Gutu District, Masvingo Province, Zimbabwe. *Climate Change Adaptations in Dryland Agriculture in Semi-Arid Areas*, 157–172.

Muzari, W., Nyamushamba, G., & Soropa, G. (2016). *Climate Change Adaptation in Zimbabwe's Agricultural Sector International Journal of Science and Research, 5*, 1762–1768.

Mwadzingeni, L., Mugandani, R., & Mafongoya, P. L. (2022). Socio-demographic, institutional and governance factors influencing adaptive capacity of smallholder irrigators in Zimbabwe. *PLoS One, 17*(8), e0273648.

Nandi, R., Pratheepa, C. M., Nedumaran, S., Rao, N., & Rengalakshmi, R. (2022). Farm parent youth aspirations on the generational succession of farming: Evidence from South India. *Frontiers in Sustainable Food Systems, 5*, 804581. https://doi.org/10.3389/fsufs.2021.804581

Nelson, S., & Huyer, S. (2016). *A gender-responsive approach to climate-smart agriculture: Evidence and guidance for practitioners. Climate-smart agriculture practice brief.* CGIAR Research Program on Climate Change, Agriculture and Food Security (CCAFS).

Nelson, S., Frankenberger, T., Langworthy, M., Finan, T., & Bower, T. (2016). The effect of livelihood diversity on recovery and shock impact in Ethiopia, Kenya and Uganda. Report prepared by the technical consortium, a project of the CGIAR. In *Technical report series no 2: Strengthening the evidence base for resilience in the horn of Africa*. A joint International Livestock Research Institute (ILRI) and TANGO International publication.

Ngenoh, E., Kebede, S., Bett, H., & Bokelmann, W. (2018). Coping with shocks and determinants among indigenous vegetable smallholder farmers in Kenya. *Agricultural Science, 9*, 804–823. https://doi.org/10.4236/as.2018.97057

Ngigi, S. N., & Denning, G. (2009). *Climate change adaptation strategies: Water resources management options for smallholder farming systems in sub-Saharan Africa*. MDG Centre.

Ngigi, M., Mueller, U., & Birner, R. (2015). *The role of livestock portfolios and group-based approaches for building resilience in the face of accelerating climate change: An asset-based panel data analysis from rural Kenya*. Available at SSRN 2676574.

RCRC (Red Cross Climate Centre). (2021). *Climate profiles of countries in southern Africa: Zimbabwe*. RCRC.

Reason, C., & Keibel, A. (2004). Tropical cyclone Eline and its unusual penetration and impacts over the southern African mainland. *Weather and forecasting., 19*(5), 789–805.

Sam, A. S., Abbas, A., Padmaja, S. S., Kaechele, H., Kumar, R., & Müller, K. (2019). Linking food security with Household's adaptive capacity and drought risk: Implications for sustainable rural development. *Social Indicators Research, 142*(1), 363–385.

Schulze, R. (2010). *Methodological approaches to assessing eco-hydrological responses to climate change in South Africa*. Water Research Commission.

Thornton, P. K., Jones, P. G., Alagarswamy, G., Andresen, J., & Herrero, M. (2010). Adapting to climate change: Agricultural system and household impacts in East Africa. *Agricultural Systems, 103*, 73–82.

Tirado, M. C., Clarke, R., Jaykus, L. A., McQuatters-Gollop, A., & Frank, J. M. (2010). Climate change and food safety: A review. *Food Research International, 43*(7), 1745–1765.

UNDP. (2020). Zimbabwe. *Climate Change Adaptation*. https://www.adaptation-undp.org/explore/easternafrica/Zimbabwe

UNDP. (2021). ZRBF resilience knowledge hub. In S. Martin, J. Stack, S. Nelson, & M. Langworthy (Eds.), *ZRBF outcome monitoring survey: Round three program learning report*.

UNDRR. (2015). *Sendai framework terminology on disaster risk reduction*. https://www.undrr.org/terminology/resilience

UNESCO-UN-Water. (2020). *United Nations world water development report 2020: Water and climate change*. UNESCO.

Vermeulen, S. J., Campbell, B. M., & Ingram, J. S. (2012). Climate change and food systems. *Annual Review of Environment and Resources, 37*(1), 195–222.

Wainaina, P., Tongruksawattana, S., & Qaim, M. (2018). Synergies between different types of agricultural Technologies in the Kenyan Small Farm Sector. *The Journal of Development Studies, 54*(11), 1974–1990. https://doi.org/10.1080/00220388.2017.1342818

WFP. (2020). World food programme. In *Zimbabwe 2020 annual country report highlights*.

World Bank. (2020). *Climate risk profile*. The World Bank Group.

Xia, J., Ning, L., Wang, Q., Chen, J., Wan, L., & Hong, S. (2017). Vulnerability of and risk to water resources in arid and semi-arid regions of West China under a scenario of climate change. *Climate Change, 144*(3), 549–563.

Yamba, F. D., Walimwipi, H., Jain, S., Zhou, P., Cuamba, B., & Mzezewa, C. (2011). Climate change/variability implications on hydroelectricity generation in the Zambezi River basin. *Mitigation and Adaptation Strategies for Global Change., 16*(6), 617–628.

Zamasiya, B., Nyikahadzoi, K., & Mukamuri, B. B. (2021). Determinants of household food security using a hierarchical linear modelling technique: A case study of smallholder farmers in Hwedza district in Zimbabwe. In K. Nyikahadzoi & B. B. Mukamuri (Eds.), *Building climate change resilient communities in Africa: Insights from Zimbabwe's urban and rural areas*. Konrad Adenauer Stiftung.

ZimVAC. (2016). Rural livelihoods assessment report.

ZimVAC. (2018). *Households and community survey*. Food and Nutrition Council Publications.

ZRBF-ECRIMS Quarterly Report. (2019). *April to June Progress Report*.

Chapter 2
Climate Smart Agriculture Practices and Gender Differentiated Food and Income Security Outcomes: Evidence from Zimbabwe

Teresa Nyika, Modester Dengedza, Carren Pindiriri, and Andrew Mupwanyiwa

2.1 Introduction

Household food and income security have become global issues among scholars and policymakers as they form an important part of human life (Agarwa et al., 2021). The United Nations Sustainable Development Goals (UN-SDGs), specifically SDG2, emphasise the need to end hunger, achieve food security and improved nutrition and promote sustainable agriculture. Achieving household food security is a crucial component of these broad developmental goals (United Nations 2015).

Millions of smallholder farmers in sub-Saharan Africa are already being negatively impacted by the frightening effects of climate change on food security and livelihoods (Ouedraogo, 2019). Currently, there are over 690 million hungry people in the world (FAO et al., 2020). Across Africa, and the world at large, notable studies have shown how climate change had severe effects on agriculture and incomes of both men and women. One of the numerous climate-related difficulties that confront farmers is rainfall variability (early start and early conclusion of the rain season, delay of the winter season, rain out of season, long dry spells, floods), which affects the water requirements of important crops (FAO et al., 2020). Furthermore, high winds which cause land erosion, kill plants and speed up the pace at which moisture evaporates from the soil. On the other hand, rising temperatures, which bring on dry spells and droughts, impact the temperature requirements of key crops and animals (FAO, 2021).

Due to inadequate rainfall and unpredictable weather patterns which negatively impacted crop harvests and prospects for a living, over 7.7 million people in Zimbabwe (almost half of the country's population) experienced food insecurity in

T. Nyika (✉) · M. Dengedza · C. Pindiriri · A. Mupwanyiwa
Department of Economics and Development, Faculty of Business Management Sciences and Economics, University of Zimbabwe, Harare, Zimbabwe

L. Mhlanga et al. (eds.), *Building Resilience to Climate Change in Selected Districts in Zimbabwe*, https://doi.org/10.1007/978-3-031-80529-5_2

2020 (WFP, 2020). The establishment of the Zimbabwe Resilience Building Fund (ZRBF) in 2015 was prompted by the country's rural communities experiencing numerous severe economic, environmental and social shocks and stresses, as well as the prevalence of climate-related hazards. UNDP manages the Fund, which is also partially funded by DFID, EU, Sweden and UNDP. The Ministry of Lands, Agriculture, Fisheries, Water, Climate and Rural Resettlement (MLAWCRR) of the Government of Zimbabwe leads the collaborative implementation of the ZRBF. The fund objectives are to promote enhanced well-being in the face of shocks and stresses, safeguard development gains and strengthen the resilience of at-risk communities (Langworthy et al., 2021). One of ZRBF's main goals is to build individuals', households', communities' and systems' resilience in times of climate shocks and stresses. To improve long-term household and community resilience, build evidence and create a crisis moderator mechanism that can give quick and flexible financing during times of shocks and stresses, the program organised activities around these three interconnected components (Langworthy et al., 2021). The ZRBF supported projects using three rounds of a household panel survey and two rounds of qualitative data. In order to monitor the effect of ZRBF-funded initiatives on 11 top-line indicators and to gather household data for annual reporting, three Outcome Monitoring Surveys (OMS) were utilised. The results of the first round showed that qualitative research was necessary to have a deeper understanding of the forces and effects behind change. This was added in round 2 and round 3. Data was collected on the 11 ZRBF top-line indicators including outcome variables and impact variables to include household income and food security, which is the focus in this study.

Intervention intensity and consortium partner were used to separate the ZRBF findings and program intervention effectiveness indicators. Collective Action for Transformation, Value Chain and Business Activities, ISALs, Climate-Smart Agricultural (CSA) Practices, Improved Livestock Practices, and Water Infrastructure and Community Gardens are the six intervention categories in which the ZRBF was active. Results from OMS round 3 show that 76% of beneficiaries participated in CSA practices, followed by improved livestock practices (69%) and value chain/business activities (61%). Among CSA practices, activities involving small grains and legumes recorded the highest levels of participation in all three OMS rounds.

According to results from the OMS round 1, almost two-thirds (65.3%) of ZRBF households reported they had experienced moderate to severe food insecurity in the previous 12 months and more than half (52.7%) reported the same in the previous 30 days. The OMS round 1 report indicated that using the Household Diet Diversity score, 50% or less of households consumed fruits, meats/fish/eggs and dairy product/milk in the last 7 days, where consumption of fruits was the lowest at less than a quarter (22%) of households (OMS1). The current level of food insecurity is supported by OMS2 results, which show that over time, households and communities have experienced more shocks, which has increased food insecurity and increased reliance on coping strategies.

Compared to OMS1 or OMS2, fewer households in OMS3 reported being exposed to drought or variable rainfall; at OMS2, this percentage was over 80 and 75% of households, respectively, whereas at OMS3 it was 49 and 21%, respectively. Furthermore, the percentage of households exposed to excessive rains increased from roughly 5–6% at OMS1 and OMS2 to 70% at OM3. Regarding resilience capacities, every index showed a notable increase between round 1 and round 2, as well as between round 2 and round 3. Between rounds 2 and 3, overall, food security and other outcomes improved, frequently reaching or exceeding their round 1 levels. Compared to earlier rounds, there was less food insecurity and less usage of destructive coping strategies in round 3.

Variations in rainfall have also lowered crop output in other regions. Additionally, data from the surveys demonstrate that households employed a variety of copying strategies in the event of shocks, and that during a drought, they resorted to ever-more-extreme or destructive coping strategies, typically beginning with a decrease in food consumption, before depleting savings, selling small livestock and selling household and productive assets. It is therefore ideal to develop a sustainable way, through CSA, of reducing these threats to food security. The ZRBF made use of several CSA practices during the 3 OMS periods which were aimed at enhancing household resilience against climate change in Zimbabwe. These include, among others, mulching, intercropping, conservation agriculture, integrated crop-livestock management and others to mention but a few. Beneficiaries of these carefully thought-out CSA practices have reported that ZRBF programming is strengthening household members' capacity for resilience and enhancing their ability to handle stress and shocks. In light of this, the purpose of this study is to learn more about how the CSA practices that ZRBF implemented have affected household food security and incomes within the project consortiums. The study will make use of data from the 11 top-line indicators used across the 3 OMS rounds.

2.2 Analytical Instruments and Sample Frame

2.2.1 Sample Design

2.2.1.1 Study Area

The research made use of data from three OMS of the ZRBF in the project areas, which are the 7 consortiums across the 18 districts in Zimbabwe.

2.2.1.2 Sampling Methods

This study adopted a mixed approach research design through the combined use of quantitative and qualitative approaches. The study adopted data which was used in the Outcome Monitoring Survey 1 to 3 for the Zimbabwe Resilience Building Fund.

The primary aim of the survey was to assess and gauge the effectiveness of the ZRBF interventions on the adaptation and resilience of beneficiary communities in Zimbabwe. The ZRBF adopted a two-stage cluster sampling design where wards were put into clusters, and these were classified into medium intensity and high intensity sampling frame. Medium intensity implied beneficiaries who had participated in two or less programme interventions and high intensity implied those who had participated in more than two interventions. To sample the clusters, the Probability Proportional to Size (PPS) approach was employed. According to the number of beneficiaries in each project district, the clusters were chosen for each project consortium across all project districts.

Sampling Frame The enumeration area was made up of the ZRBF consortiums. The sample comprised of the same households that participated in the OMS1 up to OMS3. The sample size was disaggregated based on intervention intensity, sex and consortia partners. The sampling frame comprised an equal number of medium density sampling frames and high-density sampling frames. This enabled comparison to be made across the consortiums. Table 2.1 provides the sampling frame from the 3 rounds.

2.2.1.3 Data Sources and Methods

The study made use of data collected during the OMS 1 to 3 of the ZRBF. The data was further cleaned and analysed using STATA. The following section provides the methods that were employed in this study.

2.2.2 Analysis Techniques

The study made use of comprehensive descriptive statistics and econometric methodology. First, we used various statistics such as means, variations and hypotheses tests for mean differences in income and food outcomes between men and women. These statistics provided differences between men and women in outcome variables and in the various CSA interventions (interventions in both crop and livestock production). Second, the study applied econometric techniques, in particular, panel data methods to estimate the determinants of CSA adoption and the impact of CSA adoption on food security and income. Since the same households were followed in the three waves, a panel data methodology is appropriate. Although same households were followed over 3 years, some dropped in 2021, and some were added in 2021 (see Table 2.1 which shows the number of sampled beneficiaries in the three panels). About 87 beneficiaries in the first round (2019) dropped in 2020 and 36 new respondents were added in 2021. Nevertheless, the study identified no systematic patterns of those who dropped or were added; hence, no evidence of bias from attrition. We estimated both pooled and random effects probit to resolve the problems of unbalanced panels. The *xtprobit* and *xi* : *probit* from STATA allows the estimation of unbalanced panels.

Table 2.1 Wave samples

Wave (Round)	Year	Sample size
1	2019	3440
2	2020	3353
3	2021	3389

Source: Authors' computations from data

2.2.3 Determinants of Adoption of Climate Smart Agricultural Practices

Comparisons were made to determine which climate smart agricultural practices are mostly adopted by female-headed vs male-headed households to evaluate the gender component. This enabled the research to draw the levels of resilience among men and women in the consortiums across the three periods. Since the nature of adoption measurement is binary, a limited dependent variable model becomes the most appropriate. At household level, the farmer can choose to adopt different combination of practices in each season. The choice of the CSA has implications on the household's food security outcome as well as on incomes. The adoption function is as follows:

$$CSA\,\text{Adoption}_{\text{it}} = f\left(X_{\text{it}}\right) \tag{2.1}$$

where CSA $\text{Adoption}_{\text{it}}$ is a binary variable indicating whether beneficiary i has adopted climate smart agricultural practice in time t or not, and X_{it} is a vector of the correlates of CSA adoption presented in Tables 2.2a and b. All CSA practices were measured as a binary indicator with only two categories: yes or no. Those who have used the CSA practice would respond by "yes" and those who would not have used the practice will respond by "no". These categories were similar even in the livestock practices. Characteristics of the beneficiary (age, marital status, education, household size, access to extension services, climate shock experiences, knowledge about the practices and participation in ZRBF, among others) were used as correlates of adoption. Similar correlates were applied for both CSA in crop production and livestock production. The CSA practices in livestock production are presented in Table 2.2b. The variables household head sex, ZRBF participation (ZRBF part), marital status, access to extension services and knowledge of the practice take a value of 1 if head is male, if beneficiary participates in ZRBF, if beneficiary is married, if beneficiary has access to extension services, if beneficiary has knowledge about the practice, respectively, and zero otherwise. The other correlates are continuous: age is measured in years, household size and climate shock experiences in numbers, while education is measured as categorical (no education, primary, secondary, and tertiary). The probability of adoption for each beneficiary is as follows:

$$\Pr\left(\text{adoption} = 1|\mathbf{X}\right) = \Phi\left(z\right) = \Phi\left(\mathbf{X}_{\text{it}}\beta\right) \tag{2.2}$$

Table 2.2a CSA adoption in crop production and its correlates

Dependent variables (Y): CSA practices	Independent variables (X)
Quality certified seeds	Household head sex
Community seed banks	Household size
Improved varieties	ZRBF part
Small grains	Marital status
Crop rotation	Education level
Intercropping	Climate shock
Cover cropping	Access to extension services
Mulching	Age
Integrated pest management	Knowledge of the practice
Organic fertilizer	
Drip irrigation	
Plant intensity	

Table 2.2b Selected CSA adoption in livestock production

Dependent variables (Y): CSA practices	Independent variables (X)
Animal breeds	Household head sex
Animal shelter	Household size
Water infrastructure	ZRBF part
Vaccine use	Marital status
Dozing or deworming	Education level
Dipping	Climate shock
Fodder use	Access to extension services
Survival feeding	Age
Purchasing of feedstock	Knowledge of the practice

where $\Phi(z)$ is the cumulative standard normal distribution function and **X** is a vector of explanatory variables explained above with $\Phi(z) = P(Z \leq z)$, $Z \sim N(0, 1)$.

2.2.4 *Determining the Impact of Adoption of CSA Practices on Food Security*

For the non-continuous dependent variables (food security), limited dependent variable models were applied. The study employed the panel probit model making use of an unbalanced panel dataset collected over three rounds (2019, 2020 and 2021) from beneficiaries. The same beneficiaries were followed over the three periods; however, there was some attrition where some households dropped in 2021 and new ones were added. To address the issue of attrition, the study made use of both pooled and random effects probit. The panel probit model was as follows:

$$Y_{it} = 1\left(X_{it}\beta + \alpha_i f_t + u_{it} > 0\right), i = 1, \ldots\ldots, N, t = 1, \ldots\ldots, T \quad (2.3)$$

Table 2.3 Food security and income impact model variables

Dependent variables (food security and income)	Independent variables (X)
Income	CSA practices
Beneficiaries lacking health food	Sex
Beneficiaries lacking food variety	Household size
Beneficiaries skipping meals	ZRBF part
Beneficiaries running out of food	Marital status
	Education level
	Climate shock
	Access to extension services
	Age

Where Y_{it} is the binary response variable, X_{it} is a $px1$ vector of regressors, β is a vector of parameters, f_t is a vector of unobserved time effects and α_i is a vector of unobserved individual fixed effects. The variables used for estimation of impact of CSA practices on food security and household income are presented in Table 2.3.

2.2.5 *Impact of Adoption of CSA Practices on Household Income*

Econometric models were used to establish the association between the outcome variables and CSA interventions. Regressions were done by gender to identify factors that make men and women more resilient. Specifically, the study made use of multiple regression models where the dependent variable, income, is a continuous variable as follows:

$$Y_{it} = \beta_0 + \beta_1 X_{1t} \ldots\ldots + \beta_K X_{Kt} + U_{it} \tag{2.4}$$

The study made use of other control variables which influence food security, such as education of the household head, age of the household head, access to finance, access to markets, training of CSA practices, ownership of assets, human capital (size of the household), gender of household head and other relevant variables.

To evaluate the income and food security impacts of CSA practices, there is a need to estimate the expected value of income and food and nutrition security status which adopting households would have without adoption, which is known as conditional expectation (Maddala, 1986). This was done as follows: A particular household that adopts CSA intervention has an expected value of food security which is measured by:

$$E\left(ycs|l = 1\right) \tag{2.5}$$

where ycs represents the household food security realised when one adopts CSA practices, and $l = 1$ implies that a household has adopted.

Comparisons were made between female-headed and male-headed households. The change in income and food security due to adoption of CSA technologies was determined by comparing the expected value of income/food security across the 3 periods per each household. As such the study focused again on Outcome 4, which shows average incomes across the 3 periods disaggregated by sex. This value was being measured across the sampled farmers in the consortiums, and comparisons were made to see which consortiums experienced changes in these variables. All other factors that could enhance incomes and food security were held constant so as to ensure that the changes are only attributed to CSA practices.

2.3 Data Analysis Procedures

Before the data were analysed, it was cleaned to ensure that all the data were complete and had been correctly entered. Analysis was done using the STATA software, which is well-suited and robust for analysis of all kinds of quantitative and qualitative data. In the qualitative data content, thematic analysis was used to get the common standing of the participant's views on the project interventions.

2.4 Results and Discussion

In this section, we present the findings on climate-smart agricultural practices, their uptake and impact on smallholder farmers' incomes and nutrition. These findings also show changes over time from 2019 to 2021. First, we describe CSA practices and present the factors that influence their uptake. Second, we describe income and food security outcomes and present the income and food security benefits of CSA practices to men and women. The findings are from an unbalanced panel dataset collected over three rounds (2019, 2020 and 2021) from beneficiaries.

2.4.1 CSA Practices and Determinants of Their Uptake

The results in Table 2.4 reveal that utilisation of CSA practices in crop production increased between 2019 and 2021 despite a drop in usage of a few practices (quality certified seeds, community seed banks and seed variety) in 2020. The use of quality seeds, small grains, crop rotation, improved varieties, mulching and integrated pest management are the most popular CSA practices among smallholder farmers. For instance, 84.4% of the respondents used quality certified seeds in 2021 and 81.4% grew small grains while 78.5% applied crop rotation. CSA practices such as community seed banks and drip irrigation were the least common among farmers in all three rounds of the study. Integrated pest management has also a significant

Table 2.4 Uptake of CSA practices in crop production

Practice	Percent of beneficiaries who used the practice		
	Round 1 (2019)	Round 2 (2020)	Round 3 (2021)
Quality certified seeds	79.8%	76.3%	84.4%
Community seed banks	36.5%	34.1%	43.4%
Improved varieties	75.2%	71.9%	76.6%
Small grains	71.2%	79.5%	81.4%
Crop rotation	68.7%	76.0%	78.5%
Inter-cropping	65.8%	69.8%	74.2%
Cover cropping	52.8%	66.2%	69.7%
Mulching	60.1%	62.1%	65.6%
Integrated pest management	71.7%	75.9%	78.5%
Organic fertilizer/ composite	64.7%	67.2%	66.2%
Drip irrigation	19.2%	21.4%	20.6%
Plant density	66.7%	69.6%	66.0%
Overall CSA uptake	46.8%	60.2%	72.4%

Source: Authors' computations from data

percentage of farmers practicing it despite it being an unfamiliar practice. Overall, there was a significant increase in the uptake of CSA practices by beneficiaries. For instance, the proportion of beneficiaries making use of CSA practices rose from 46.8% during the first round in 2019 to 72.4% during the third round in 2021. This is supported by the panel probit findings, which show a statistically significant positive coefficient of year 2021 or round three in the CSA uptake model in Table 2.5. Similarly, the proportion of beneficiaries making use of CSA in livestock production increased between 2019 and 2021, except in the use of produced fodder and purchased feed stock (see Table 2.5).

The dataset applied in this chapter collected by the Zimbabwe Resilience Building Fund is biased towards female participants. Figure 2.1 shows that in the first round (2019), 58% of household heads were female participants while 42% were male participants; however, in the second and third rounds (2020 and 2021), only 31% and 32%, respectively, were male. During the first round in 2019, there were no significant differences in the proportion of female and male CSA adopters in all the crop production CSA practices, except in drip irrigation where there was a significantly larger percentage of households headed by males applying it. In livestock production, there was a larger percentage of households headed by males utilizing improved animal breeds, water infrastructure and fodder production compared to females. For the other CSA practices in livestock production, there were no significant differences between male- and female-headed households, except in dozing and dipping. From 2020, significant differences existed between males and females in the adoption of CSA practices. The proportion of women adopting these practices started to be significantly larger than that of males (see Table 2.6). It is

Table 2.5 Uptake of selected CSA practices in livestock production

Practice	Percent of beneficiaries who used the practice		
	Round 1 (2019)	Round 2 (2020)	Round 3 (2021)
Improved animal breeds	29.1%	36.9%	39.5%
Improved animal shelter	41.1%	48.4%	48.5%
Improved water infrastructure	34.7%	47.9%	40.2%
Routine vaccination	29.1%	36.9%	39.5%
Dozing application	56.0%	60.6%	60.7%
Dipping	65.2%	71.9%	74.9%
Produced fodder	42.0%	38.6%	41.3%
Survival feeding	50.8%	57.7%	55.5%
Purchased feed	33.2%	35.1%	34.3%

Source: Authors' computations from data

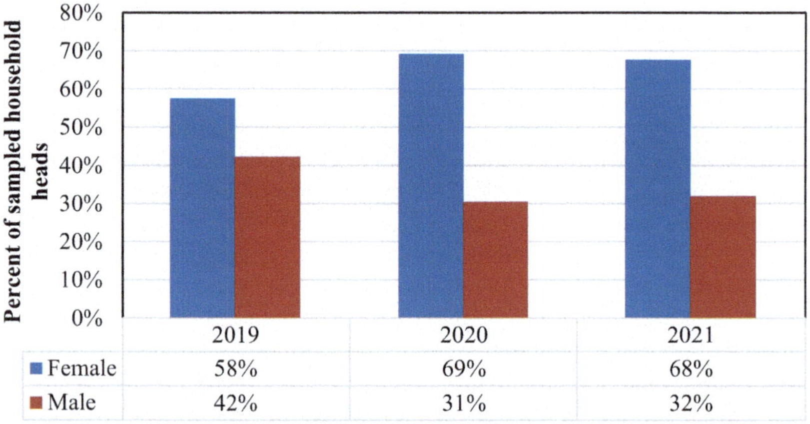

Fig. 2.1 Share of women in sampled beneficiaries. (Source: Authors' illustration)

however crucial to treat the findings on gender differences with caution as targeting women in beneficiation introduces gender self-selection bias in the findings.

The mean age of beneficiaries in the ZBRF projects is 47 years. The results, however, demonstrate that there is no evidence for age group differential in the usage of CSA practices. The other variables correlated with CSA practices are in the panel probit regressions in Table 2.7. These variables include household size, education of members in the household, education of participants, marital status of participants, participation in ZBRF projects and climate change related shocks. An increase in the frequency of climate shocks increases the probability of adopting

Table 2.6 Gender-differentiated proportion of household heads using CSA practices

Characteristic (mean prop)	2019			2020			2021		
Crop production	Female	Male	Difference	Female	Male	Difference	Female	Male	Difference
Quality certified seeds	81.8%	81.6%	0.2	80.3%	72.7%	7.6***	88.4%	82.6%	5.8***
Community seed banks	36.4%	34.5%	1.9	34.2%	28.3%	5.9**	41.8%	42.9%	−1.1
Improved varieties	77.2%	77.0%	0.2	75.7%	68.8%	6.9***	81.1%	75.9%	5.3***
Small grains	74.8%	75.7%	−0.8	82.6%	78.9%	3.7***	86.1%	83.1%	3.0**
Crop rotation	72.1%	74.3%	−2.1	80.6%	74.4%	6.2***	84.7%	77.8%	6.9***
Inter-cropping	69.9%	68.7%	1.1	71.3%	70.6%	0.7	79.5%	75.3%	4.1***
Cover cropping	55.1%	57.9%	−2.8	66.0%	66.4%	0.4	73.4%	69.75%	3.7**
Mulching	64.5%	64.2%	0.3	64.5%	61.9%	2.6	71.3%	62.2%	5.1***
Pest management	74.4%	74.4%	0.0	78.8%	72.4%	6.4***	82.7%	77.7%	4.1***
Organic fertilizer	65.9%	65.3%	0.6	68.2%	63.9%	4.3**	70.3%	61.8%	8.6***
Drip irrigation	19.1%	28.8%	−9.7***	21.1%	24.9%	−3.7*	21.8%	22.5%	−0.7
Plant density	68.8%	71.1%	2.4	72.6%	68.4%	4.2**	72.8%	64.2%	8.6***
Livestock production									
Animal breeds	30.6%	36.7%	−6.1***	41.8%	33.9%	8.0***	44.3%	37.7%	6.5***
Animal shelter	45.0%	43.3%	1.7	52.5%	46.4%	6.1***	53.7%	47.4%	6.2***
Water infrastructure	37.5%	41.1%	−3.6*	48.8%	43.7%	5.1**	44.4%	38.7%	5.7***
Vaccination	64.2%	66.1%	−1.9	68.8%	58.8%	10***	66.2%	58.7%	7.5***
Dozing	58.7%	53.9%	4.9***	63.7%	51.5%	12.3***	66.6%	51.3%	15.3***
Dipping	67.6%	64.4%	3.2**	77.2%	63.4%	13.8***	81.0%	67.2%	13.8***
Fodder production	46.5%	50.5%	−4.0*	43.9%	39.5%	4.4*	47.2%	41.5%	5.7**
Survival feeding	56.1%	55.0%	1.1	61.8%	53.3%	8.6***	62.4%	53.4%	9.0***
Feed purchases	36.0%	36.8%	−0.8	37.7%	32.4%	5.3**	40.6%	34.9%	5.7***

***, ** and * indicate that the difference between the proportion of female head CSA adopters and male head adopters is statistically significant at 1%, 5% and 10%, respectively

Table 2.7 Correlates of CSA uptake from panel probit model

	(1)	(2)	(3)	(4)	(5)	(6)	(7)	(8)	(9)	(10)	(11)
Variables	CSA uptake	Quality seeds	Comm seed banks	Small grains	Crop rotation	Inter-cropping	Cover cropping	Mulching	Integrated pest management	Organic fertilizer/ composite	Drip irrigation
Hhsex	−0.169**	−0.210***	−0.124	−0.124	−0.376***	−0.277***	−0.200***	−0.248***	−0.186***	−0.278***	−0.173***
	(0.0703)	(0.0541)	(0.0947)	(0.0947)	(0.0514)	(0.0598)	(0.0666)	(0.0539)	(0.0582)	(0.0538)	(0.0631)
Hhsize	0.0243**	0.0224*	0.0138	0.0138	0.0407***	0.0336**	−0.00209	0.00857	0.00112	−0.000727	0.00346
	(0.0104)	(0.0117)	(0.0192)	(0.0192)	(0.0114)	(0.0133)	(0.0141)	(0.0116)	(0.0121)	(0.0115)	(0.0127)
ZRBF part	0.789***	0.589***	0.332***	0.332***	0.543***	0.539***	0.446***	0.487***	0.396***	0.493***	0.394***
	(0.0502)	(0.0551)	(0.101)	(0.101)	(0.0520)	(0.0612)	(0.0718)	(0.0554)	(0.0599)	(0.0553)	(0.0688)
Mstatus	0.0754	0.162**	0.102	0.102	0.189***	0.192***	0.104	0.103	0.190***	0.0944	0.0917
	(0.0751)	(0.0641)	(0.111)	(0.111)	(0.0612)	(0.0709)	(0.0804)	(0.0641)	(0.0686)	(0.0642)	(0.0759)
Peduc	0.115***	0.144***	0.00861	0.00861	0.0808***	0.0675**	0.102***	0.0974***	0.111***	0.163***	0.0774***
	(0.0225)	(0.0263)	(0.0408)	(0.0408)	(0.0246)	(0.0275)	(0.0305)	(0.0251)	(0.0269)	(0.0260)	(0.0278)
Climate shock	0.111***	0.0897***	0.182***	0.182***	0.0747***	0.0808***	0.113***	0.107***	0.0902***	0.128***	0.122***
	(0.0170)	(0.0192)	(0.0354)	(0.0354)	(0.0186)	(0.0212)	(0.0242)	(0.0196)	(0.0209)	(0.0197)	(0.0222)
Acc_ext	1.076***	0.942***	0.737***	0.737***	0.903***	0.997***	0.886***	0.828***	0.880***	0.891***	0.752***
	(0.0596)	(0.0677)	(0.145)	(0.145)	(0.0610)	(0.0780)	(0.0933)	(0.0692)	(0.0736)	(0.0665)	(0.0962)
Age	−0.00604***	−0.00337**	−0.006**	−0.00616**	−0.00199	−0.0064***	−0.00467**	−0.00222	−0.00281	−0.00215	−0.00275
	(0.00147)	(0.00166)	(0.00286)	(0.00286)	(0.00155)	(0.00183)	(0.00210)	(0.00164)	(0.00177)	(0.00163)	(0.00194)
2. Primary	0.0825	0.0621	0.200	0.200	−0.0565	−0.0925	0.0695	−0.0427	0.0142	−0.0589	0.0499
	(0.0661)	(0.0728)	(0.128)	(0.128)	(0.0712)	(0.0839)	(0.0944)	(0.0759)	(0.0807)	(0.0758)	(0.0919)
3. Secondary	−0.0129	0.131	0.245	0.245	−0.0219	0.00345	0.0175	0.0161	0.0668	−0.0727	0.144
	(0.0761)	(0.0855)	(0.150)	(0.150)	(0.0825)	(0.0965)	(0.106)	(0.0859)	(0.0927)	(0.0862)	(0.101)

4. Tertiary	−0.174	−0.00353	0.290	0.290	0.0428	0.0972	−0.0488	−0.0606	−0.142	0.152	0.0658
	(0.153)	(0.171)	(0.266)	(0.266)	(0.177)	(0.192)	(0.204)	(0.167)	(0.170)	(0.188)	(0.178)
Constant	−1.013***	−0.621***	−0.460*	−0.460*	−0.416***	−0.299*	−0.270	−0.271*	−0.277*	−0.431***	−0.893***
	(0.147)	(0.149)	(0.276)	(0.276)	(0.143)	(0.169)	(0.193)	(0.151)	(0.165)	(0.151)	(0.191)
/lnsig2u	−1.083	−1.5968	−0.6177	−1.8429	−1.2141	−1.1372	−1.3472	−1.9602	−11.656	−1.4189	−1.7169
sigma_u	0.582	0.4500	0.7343	0.3979	0.5449	0.5663	0.5099	0.3753	0.0029	0.4919	0.4238
Rho	0. 0.254	0. 1684	0.3503	0.1367	0.2290	0.2428	0.2063	0.1234	0	0.1948	0.1523
chibar2(01)	45.2***	9.6***	10.4***	7.6***	19.2***	16.5***	5.9***	4.4**	0.0001	13.3***	4.0**

*, ** and *** indicate statistically significant at 1%, 5% and 10%, respectively. Chibar2 tests goodness of fit

CSA practices. Climate change related shocks include excessive rains, floods, variable rains, droughts, pests, reduced soil fertility, deforestation and frost. In regions where there is high frequency of climate related shocks, we expect higher uptake of CSA practices as revealed by the findings. Although gender issues are considered with caution as explained earlier in this section, the findings show that male participants have a lower probability of adopting CSA practices. Participants of ZRBF have a higher probability of adopting CSA practices. The other factors that promote uptake of CSA practices include education and access to extension services. Programs that improve rural education and mobility of agricultural extension officers can significantly promote the uptake of CSA practices. The regression results in Table 2.7 are also supplemented by Table 2.7, which shows that uptake of CSA practices was significantly higher in the third round (2021) compared to the first round of 2019.

The model of CSA uptake correlates is presented in Table 2.7 using different estimation techniques. The findings show that the main correlates of CSA uptake are sex, education, frequency of climate shocks, household size, access to agricultural services, age, marital status and participation in ZRBF projects. These variables continue to be significant in influencing the probability of uptake of some individual CSA practices irrespective of the use of a different estimator. Please note that the panel probit findings in Tables 2.7 and 2.8 only suggest the relationship between the uptake of CSA practices and the given factors. The results show that education (both the number of adults with at least primary education in a household and the participant having at least primary education), climate shock frequency, participating in ZBRF projects and access to agricultural information promote the uptake of CSA practices. The statistically significant positive coefficient of the year 2021 buttresses the findings that the probability of CSA practices uptake increased significantly in 2021 from 2019.

In Table 2.9 adoption is compared among districts. The other districts were dropped because of inadequate data from a full sample. The first model in Table 2.9 compares Mudzi's adoption of CSA practices with other districts. Mudzi is the reference district in Model 1, Chiredzi in model 2, Zvishavane in model 3, Umguza in model 4, Nyanga in model 5 and Matobo in model 6. The way reference districts were selected was to have at least one reference district from these different regions. The findings show that adoption of CSA practices is lower in the Matebeleland districts compared to Mashonaland regions. CSA adoption is highest in Masvingo and Midlands districts such as Chiredzi, Mberengwa and Zvishavane. Compared to Umguza, Nyanga and Matobo, Masvingo and Midlands districts have higher adoption levels, and the differences are statistically significant. However, comparing Mudzi and Masvingo-Midlands districts, the CSA adoption differences are statistically insignificant.

Table 2.8 Panel probit models of CSA uptake correlates from different panel estimators

	(1)	(2)	(3)
Variables	CSA uptake (year dummies)	CSA uptake (pooled estimates)	CSA uptake (random effects)
Hhsex	−0.510***	−0.469***	−0.521***
	(0.0489)	(0.0409)	(0.0482)
Hhsize	0.0242**	0.0187**	0.0213**
	(0.0105)	(0.00880)	(0.0103)
ZRBF part	0.826***	0.749***	0.845***
	(0.0506)	(0.0409)	(0.0501)
Mstatus	0.344***	0.310***	0.341***
	(0.0577)	(0.0483)	(0.0567)
Peduc	0.111***	0.100***	0.111***
	(0.0227)	(0.0192)	(0.0223)
Climate shock	0.118***	0.0940***	0.105***
	(0.0175)	(0.0148)	(0.0169)
Acc_ext	1.011***	0.921***	1.030***
	(0.0603)	(0.0491)	(0.0595)
Age	−0.00542***	−0.00449***	−0.00518***
	(0.00147)	(0.00124)	(0.00144)
2. Primary	0.0767	0.0526	0.0610
	(0.0667)	(0.0574)	(0.0657)
3. Secondary	−0.0105	−0.0252	−0.0298
	(0.0766)	(0.0654)	(0.0754)
4. Tertiary	−0.200	−0.174	−0.212
	(0.153)	(0.129)	(0.151)
Year 2021	0.294***		
	(0.0423)		
Constant	−1.232***	–	−1.071***
	(0.138)	(0.114)	(0.133)
/lnsig2u	−1.0980		−1.0980
sigma_u	0.5775		0.5775
Rho	0.2501		0.2501
chibar2(01)	41.8***	Pseudo R2 = 0.17	41.8***

*, ** and *** indicate statistically significant at 1%, 5% and 10%, respectively

2.4.2 *Impact of CSA Practices on Incomes and Food Security*

The descriptive results in Table 2.10 demonstrate that the mean monthly household income of beneficiaries rose from $79 during the baseline (2019 first round) to about $129.4 by 2021. The mean difference of about $50 dollars is statistically significant (see Table 2.11). Incomes increased as the uptake of CSA practices increased during the same period. A worrying condition demonstrated in Table 2.10 is the huge variations in incomes. The coefficient of income variation is greater than 1 in

Table 2.9 District differentials

	(1)	(2)	(3)	(4)	(5)	(6)
Variables	CSA adoption (reference is Mudzi)	CSA adoption (reference is Chiredzi)	CSA adoption (reference is Zvishavane)	CSA adoption (reference is Umguza)	CSA adoption (reference is Nyanga)	CSA adoption (reference is Matobo)
Mutoko	0.355	0.108	0.180	2.792***	1.197***	1.615***
	(0.385)	(0.371)	(0.511)	(0.738)	(0.396)	(0.461)
Chiredzi	0.247		0.0719	2.684***	1.089***	1.507***
	(0.355)		(0.488)	(0.722)	(0.366)	(0.436)
Mwenezi	0.524	0.277	0.349	2.961***	1.366***	1.784***
	(0.460)	(0.448)	(0.569)	(0.779)	(0.469)	(0.525)
Mberengwa	0.358	0.111	0.183	2.796***	1.200***	1.619***
	(0.385)	(0.371)	(0.511)	(0.738)	(0.396)	(0.461)
Zvishavane	0.175	−0.0719		2.612***	1.017**	1.436**
	(0.500)	(0.488)		(0.803)	(0.508)	(0.560)
Nkayi	−1.027**	−1.274***	−1.202**	1.411*	−0.185	0.234
	(0.465)	(0.453)	(0.573)	(0.782)	(0.474)	(0.529)
Umguza	−2.437***	−2.684***	−2.612***		−1.595**	−1.177
	(0.730)	(0.722)	(0.803)		(0.736)	(0.772)
Beitbridge	−1.606***	−1.853***	−1.781***	0.832	−0.764*	−0.345
	(0.409)	(0.396)	(0.529)	(0.751)	(0.419)	(0.481)
Nyanga	−0.842**	−1.089***	−1.017**	1.595**		0.418
	(0.381)	(0.366)	(0.508)	(0.736)		(0.458)
Insiza	−1.158**	−1.405***	−1.333**	1.279	−0.316	0.102
	(0.481)	(0.470)	(0.587)	(0.792)	(0.490)	(0.544)
Lupane	−1.088**	−1.335***	−1.263**	1.349*	−0.246	0.172
	(0.429)	(0.417)	(0.545)	(0.762)	(0.439)	(0.499)
Matobo	−1.260***	−1.507***	−1.436**	1.177	−0.418	
	(0.448)	(0.436)	(0.560)	(0.772)	(0.458)	
Mbire	−0.795**	−1.042***	−0.971*	1.642**	0.0466	0.465
	(0.401)	(0.387)	(0.523)	(0.746)	(0.412)	(0.475)
Mudzi		−0.247	−0.175	2.437***	0.842**	1.260***
		(0.355)	(0.500)	(0.730)	(0.381)	(0.448)
Constant	1.763***	2.010***	1.938***	−0.674	0.921***	0.502
	(0.261)	(0.240)	(0.426)	(0.681)	(0.277)	(0.364)
Observations	700	700	700	700	700	700

Standard errors in parentheses
*** $p < 0.01$, ** $p < 0.05$, * $p < 0.1$

all three rounds (2019, 2020 and 2021), although it is smallest in 2021. However, the decline in the coefficient of variation from 3 to 1.76 in an environment with a significant increase in incomes is plausible despite high levels of incomes variability. The high-income variability coefficient findings point to the importance of

Table 2.10 Mean monthly household income

Year	Mean ($)	Standard deviation	Coefficient of variation	Observations
2019	79.0	161.6	2.05	3200
2020	86.7	260.4	3.00	3089
2021	129.4	227.7	1.76	3106
Total	**98.2**	**220.9**	**2.25**	**9395**

Source: Authors' computations from data

Table 2.11 Income and food security status differences between baseline 2019 and endline 2021

Characteristic (mean, prop)	Baseline 2019 (N = 3200)	Endline 2021 (N = 3106)	Difference	z-statistic
Income	$79.0	$129.4	-$50.4	−10.2***
Beneficiaries lacking health food	70.5%	72.3%	−1.9%	−1.7**
Beneficiaries lacking food variety	71.9%	74.9%	−3.0%	−2.8***
Beneficiaries skipping meals	58.9%	60.4%	−1.5%	−1.3*
Beneficiaries running out of food	48.5%	53.6%	−5.1%	−4.4***

Source: Authors' computations
***, ** and * indicate that the difference between the baseline and endline average is statistically significant at 1%, 5% and 10%, respectively. Difference in means was tested using t-tests for equality of means and Levene's test for equality of variances

income redistribution strategies when implementing projects such as the ZRBF. Table 2.11 provides statistical tests to test whether income and food security indicators in 2021 changed from the baseline values of 2019. The results show that food security deteriorated significantly by 2021. The proportion of beneficiaries lacking health food, food variety, skipping meals and running out of food increased in 2021 from the 2019 numbers. This may be a result of the Covid-19 pandemic and macroeconomic instability characterised by high food prices.

While the findings in Table 2.11 reveal that the proportion of beneficiaries who are food insecure increased in 2021 from the 2019 levels, it is crucial to note that the proportion of CSA adopters who are food insecure is smaller than that of non-adopters. The results in Table 2.12 show that adopters of CSA practices have a larger income and a smaller proportion of population who lack health food, lack food variety, skip meals, and run out of food. Although food security generally deteriorated between 2019 and 2021, the impact was lower for adopters of CSA practices than non-adopters. For instance, the proportion of CSA adopters who lacked

Table 2.12 Income and food security differentials between CSA adopters and non-adopters

Characteristic (Mean prop)	2019			2020			2021		
	Adopters (1627)	Non-adopters (1813)	Difference	Adopters (1627)	Non-adopters (1813)	Difference	Adopters (1627)	Non-adopters (1813)	Difference
Income	$88.3	$68.2	20.1***	$91.5	$76.4	15.1*	$138.6	$96.1	42.5***
Lack of health food	68.1%	73.1%	−5.1***	80.5%	82.2%	−1.7	70.4%	78.8%	−8.5***
Lack of food variety	69.1%	75.1%	−6.1***	82.6%	83.5%	−0.9	73.3%	80.5%	−7.2***
Skipped meals	55.2%	62.9%	−7.2***	69.4%	75.7%	−6.4***	58.8%	65.6%	−6.8***
Ran out of food	45.4%	52.0%	−6.6***	59.9%	68.1%	−8.2***	51.3%	61.2%	−9.9***

Source: Authors' computations
***, ** and * indicate that the difference between the proportion of CSA adopters and non-adopters is statistically significant at 1%, 5% and 10%, respectively. Difference in means was tested using t-tests for equality of means and Levene's test for equality of variances with population weights applied

Table 2.13 Food security status differences between CSA adopters and non-adopters using pooled estimators

Status of food security over the last 12 months	Proportion of adopters (%)	Proportion of non-adopters (%)	Overall proportion (%)	Difference	z-statistic
Lack of health food	63.6	84.4	64.8	−20.9	−2.84***
Lack of food variety	68.1	82.2	69.0	−14.1	−1.99**
Skipped meals	54.1	64.4	54.7	−10.4	−1.35*
Ran out of food	45.0	64.4	46.2	−19.4	−2.53***

Source: Authors' computations
***, ** and * indicate that the difference between the proportion of CSA adopters and non-adopters who are food insecure is statistically significant at 1%, 5% and 10%, respectively. Difference in means was tested using z-tests for equality of means and Levene's test for equality of variances with population weights applied

health food was 8.5% lower than that of non-adopters. Similarly, the proportions of CSA adopters who ran out of food, lacked food variety and skipped meals are 9.9%, 7.2% and 6.8% lower than those of non-adopters, respectively. The same results are obtained if a pooled estimator is applied as in Table 2.13. The pooled estimators still show that the percentage of CSA adopters who are food insecure is smaller than that of non-adopters (Table 2.13). The implication of these findings is that CSA practices can lessen the impacts of natural disasters such as Covid-19 pandemic and extreme weather events on incomes and food security among the rural population. They are therefore a crucial strategy for building resilience in a changing climate with high risk of extreme weather events.

The findings presented in Table 2.14 show that adoption of CSA practices by smallholder farmers may not necessarily generate a pronounced income difference between adopters and non-adopters. But it rather creates a pronounced difference in food security. Model 1 in Table 2.14 presents the determinants of income, including CSA adoption. The findings show no evidence of the income differences between CSA adopters and non-adopters. The difference is statistically insignificant. From the models of food security (models 2 to 5 in Table 3.11), there is strong statistical evidence that adopters of CSA practices are more food secure than non-adopters. For instance, in model 2, the probability of being unable to purchase health food is lower for adopters. Similarly, the probabilities of having lack of food variety in model 3, skipping meals in model 4 and running out of food in model 5 are lower for CSA adopters than non-adopters. While the adoption of CSA practices improves food security among smallholder farmers, climate shocks are a threat to food security. An increase in the number of climate shocks increases the probability of being food insecure (see models 2 to 5 in Table 2.14). The other correlates of income and

Table 2.14 Panel regressions of the impact of CSA adoption on income and food security

	(1)	(2)	(3)	(4)	(5)
Variables	Income	Lack of health food	Lack of food variety	Skipped meals	Ran out of food
Sex	−2.530	0.0106	−0.0961**	−0.0189	0.0568
	(7.483)	(0.0473)	(0.0486)	(0.0435)	(0.0434)
Hhsize	4.519***	0.0286***	0.0496***	0.0293***	0.0331***
	(1.552)	(0.0102)	(0.0108)	(0.00940)	(0.00932)
ZRBF part	20.20***	−0.0694	−0.0770	−0.123***	−0.0499
	(7.842)	(0.0499)	(0.0517)	(0.0457)	(0.0451)
Mstatus	25.62***	−0.167***	−0.215***	−0.158***	−0.168***
	(8.765)	(0.0581)	(0.0608)	(0.0527)	(0.0524)
Peduc	7.519**	−0.0945***	−0.0919***	−0.0618***	−0.0561***
	(3.244)	(0.0208)	(0.0218)	(0.0192)	(0.0192)
CSA adoption	8.881	−0.151***	−0.210***	−0.223***	−0.230***
	(8.588)	(0.0550)	(0.0571)	(0.0502)	(0.0492)
Climate shock	4.548*	0.124***	0.153***	0.129***	0.114***
	(2.604)	(0.0164)	(0.0172)	(0.0152)	(0.0151)
Acc_ext	−4.691	−0.0925	0.0552	0.0217	−0.0622
	(10.19)	(0.0655)	(0.0665)	(0.0585)	(0.0578)
Age	−0.310	0.00157	0.00138	0.00006	−0.00124
	(0.224)	(0.00143)	(0.00147)	(0.00131)	(0.00130)
2. Primary	3.555	0.0448	0.132*	0.0488	−0.0371
	(10.64)	(0.0687)	(0.0705)	(0.0618)	(0.0614)
3. Secondary	29.71**	−0.0378	0.0390	−0.0509	−0.245***
	(11.80)	(0.0757)	(0.0780)	(0.0693)	(0.0689)
4. Tertiary	187.1***	−0.901***	−0.604***	−0.695***	−0.930***
	(22.72)	(0.139)	(0.141)	(0.136)	(0.142)
Year 2021	44.25***	−0.340***	−0.327***	−0.343***	−0.257***
	(6.833)	(0.0412)	(0.0426)	(0.0375)	(0.0367)
Constant	−2.131	1.028***	0.858***	0.620***	0.525***
	(21.12)	(0.136)	(0.139)	(0.124)	(0.123)
/lnsig2u		−1.0306	−0.9300	−1.1359	−1.0087
sigma_u	16.2	0.5973	0.6281	0.5667	0.6039
Rho	0.004	0.2630	0.2829	0.2431	0.2672
chibar2(01)	228.2***	51.7***	55.2***	55.6***	72.7***

*, ** and *** indicate statistically significant at 1%, 5% and 10%, respectively. The chibar2 for the income model is a Wald statistic

food security include education and being married which increase income and enhance food security. The results further buttress the earlier findings that the year 2021 was associated with a higher mean income level and less food insecurity for CSA adopters than non-adopters.

2.5 Summary of the Findings

First, we asked the question: What factors influence adoption of CSA practices and their gender implications? The findings generally demonstrate that the ZRBF has been playing a key role in enhancing adoption of CSA practices. Smallholder farmers participating in ZRBF projects have higher chances of adopting CSA technologies and practices. In addition to the role being played by the ZRBF, factors such as education, and the provision of agricultural information play a critical role in enhancing adoption of CSA practices. The results further show that adoption of CSA practices vary from district to district and that climate related shocks also drive smallholder farmers to adopt these CSA practices. The main gender implication from the findings is that women have a higher probability of adopting CSA practices and there are statistically significant differences with men in many CSA practices. The reason why women are more likely to adopt these practices than men may be that more women participate in ZRBF projects.

Second, we asked whether CSA practices improve smallholder farmers' incomes and food security. Although food security has generally declined between 2019 and 2021, the findings show evidence that adopters of CSA practices have higher monthly incomes and are more food secure than non-adopters. In fact, there is statistically significant difference between the mean income of adopters and that of non-adopters. On food security impacts, the results show that adoption of CSA practices has some positive returns in food security. Non-adopters of CSA practices are more food insecure than adopters. In all dimensions (lack of health food, skipping meals, lack of variety and running out of food), there is a larger percentage of non-adopters of CSA than adopters. The crucial implication of the findings is that CSA practices are critical in building resilience in a changing climate. These practices not only lessen the impacts of climate change-induced extreme weather events, but they also help poor farmers to lessen impacts of food price instability caused by other natural disasters such the Covid-19 pandemic.

2.6 Recommendations

The government and developmental organizations, or non-governmental organizations, should promote dissemination of relevant agricultural information to communities, particularly on climate change related aspects to promote adoption of CSA practices to enhance resilience building, improve household food and income security. In this regard, the role of extension education is critical. The higher probability of adoption of CSA by women require that their needs be taken into consideration at the design stage of CSA policies, programs and projects to avoid disrupting their food and income security outcomes. Overall, effectiveness of policies, programs or projects that target promotion of CSA practices must not be measured only based on their income impacts. Instead, they must also be measured based on their impact on food security and other outcomes. Using income only to measure the effectiveness of CSA practices may render them ineffective by missing the critical role they play in enhancing food security in a changing climate. We also recommend considering income redistribution strategies when implementing projects such as the ZRBF, so as not to weaken beneficiaries' resilience capacity to adverse weather conditions, pandemics such as Covid-19 and other extreme events including macroeconomic price instability.

Chapter 3
Determinants of Household Food Security Among Smallholder Farmers Under Changing Climates and Contexts: A Case of the Zambezi Valley in Zimbabwe

Byron Zamasiya, Prosper Tondera Mataruse, and Kefasi Nyikahadzoi

3.1 Introduction

Environmental, economic, and social shocks and stresses adversely affect rainfed-fed agricultural production, purchasing power and, most importantly, a household's food security (Sam et al., 2018). Southern Africa experiences shocks and stresses, where rainfed agriculture constitutes over 95% of agricultural production and remains a significant source of livelihood for millions of smallholder farmers. In this region, population growth is outstripping per capita agricultural production, and over 30% of the population is already undernourished (Food and Agriculture Organization-FAO, 2015). The Intergovernmental Panel on Climate Change (IPCC)'s fifth Report estimates that if Green House Gas emissions continue to rise unabated under a business-as-usual scenario, the average temperature for Africa will increase by 2.6–4.8 degrees Celsius by the end of the twenty-first century (IPCC, 2014). The projected increase in temperature and the proliferation of other economic and social extremes, such as COVID-19, implies that vulnerability will worsen for marginalised communities struggling to manage other shocks and stressors. Successive years of droughts, economic deterioration and social shocks increase the vulnerability of marginalised households (Shiferaw et al., 2012; Koomson et al., 2020). These shocks and stresses limit the ability of marginalised communities to generate income and access markets, erode savings and force the disposal of productive assets to finance food security requirements. The increased vulnerability weakens the marginalised communities' capacity to respond to future shocks and

B. Zamasiya (✉) · K. Nyikahadzoi
Environment, Climate and Sustainable Development Institute, University of Zimbabwe, Harare, Zimbabwe

P. T. Mataruse
Department of Community and Social Development, University of Zimbabwe, Harare, Zimbabwe

L. Mhlanga et al. (eds.), *Building Resilience to Climate Change in Selected Districts in Zimbabwe*, https://doi.org/10.1007/978-3-031-80529-5_3

stresses and worsens food security. Frankenberger et al. (2013), as they wrote earlier, note that for marginalised communities to withstand economic, environmental, and social shocks and stresses, development projects should "resiliencify". The focus for smallholder farmers should be on investing in adaptive, absorptive, and transformative capacities (Becchetti & Castriota, 2011; Bene et al., 2012). Although lean season support programmes and social cash transfers are crucial in plugging short-term household food insecurity challenges, they cannot improve the resilience of households to future shocks and stresses.

To address the vulnerability of rural farmers' communities to severe economic, social and environmental shocks in Zimbabwe, the Government of Zimbabwe, in partnership with UNDP, DFID, the Embassy of Sweden and the Government of Denmark, implemented the Zimbabwe Resilience Building Fund (ZRBF) program (Government of Zimbabwe-GoZ, 2020). The ZRBF program sought to improve the resilience of about 830,000 people through the generation and use of evidence to inform programming and policy (United Nations Development Program-UNDP, 2019a). The focus of the ZRBF program was to build the capacities of rural farmers to absorb, adapt to and transform livelihoods and systems to reduce risks at individual, household, community, ward, district and provincial levels. In this project, absorptive capacity is defined as the ability to minimise exposure to future shocks and stresses and recover quickly when exposed (UNDP, 2019a, b). Adaptive capacity involves making proactive and informed choices about alternative livelihood strategies based on changing conditions (UNDP, 2021). Transformative capacity involves the governance mechanisms, policies, regulations, infrastructure, community networks and formal and informal social protection mechanisms that constitute an enabling environment for systemic change.

The ZRBF program used a crisis modifier concept whose role was to activate an early action in response to an early warning of a crisis to protect the development gains of the program (UNDP, 2020). The ZRBF program aimed to contribute to the increased capacity of at-risk communities to protect development gains and achieve improved well-being in the face of shocks and stresses. ZRBF's focus was to build the resilience of individuals, households, communities, and systems. To achieve this goal, ZRBF organised activities around three interconnected components: building evidence, supporting long-term household and community resilience and a crisis modifier mechanism which can provide agile and flexible funding in times of shocks and stresses (UNDP, 2020). The program was implemented by a team of 7 consortia in 18 Rural District Councils of Zimbabwe from 2016–2022 (UNDP, 2019a, b). The Resilience Knowledge Hub (RKH), led by Mercy Corps Zimbabwe, worked with consortia members to support the building and nurturing of professional relationships and social capital across ZRBF partners and between ZRBF partners and external stakeholders. The RKH facilitated the building of an effective coalition for change, bringing together a strategic, multi-disciplinary team of actors to learn collectively, interact and strategise towards a more resilient Zimbabwe. The consortia were led by Action Aid Zimbabwe, Christian Aid, Danish Church Aid, Welthunger, Care International and International Rescue Committee. The program aimed to build the resilience of marginalised communities with a particular focus on adaptive, absorptive, and transformative capacities. Although resilience building looks at three capacities, this chapter focuses on how adaptive capacity drives household food security. The objective of adaptive capacity is to strengthen households' ability to make informed and proactive decisions about alternative livelihood choices, such as preparing for climate-related risks to their productive activities (UNDP, 2021). Although the consortia supported different initiatives or practices, they all aimed to strengthen the adaptive capacities of individuals, households and communities based on an understanding of changing conditions.

3.1.1 The Resilience Pathway Model

Figure 3.1 depicts the resilience pathway. According to the resilience pathway framework, strengthened resilience capacity drives household responses to shocks and stresses better, which leads to better well-being outcomes. The conceptual model posits that increased programme participation drives this pathway, i.e. program participation leads to improvements in resilience capacity, which leads to better responses by households, which in turn helps them maintain or improve their food security or other well-being outcomes in the face of shocks or stresses.

Shocks Faced by Smallholder Farmers in the Zambezi Valley Landscape
The Zambezi Valley Alliance is one of the ZRBF consortia led by Action Aid Zimbabwe and was working in Binga, Kariba and Mbire districts. Over the years, numerous shocks and stresses have affected the districts where the Zambezi Valley Alliance was working. These shocks had adverse effects on the well-being of smallholder farmers. The shocks included physical and socioeconomic shocks.

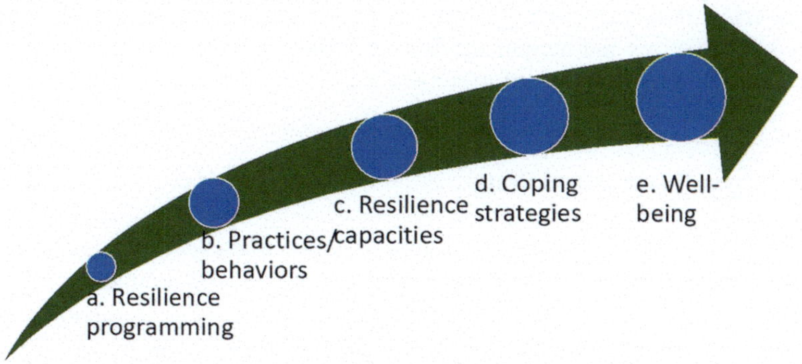

Fig. 3.1 Resilience pathway. (Source: OMS Round 3 Report UNDP, 2021)

3.1.2 *Physical Shocks*

3.1.2.1 Droughts

Since 1992, Zimbabwe has been experiencing meteorological droughts every ten years, with the worst droughts experienced in 1982 and 1992 (GoZ & UNCT Zimbabwe, 2014). Since 2000, meteorological droughts have increased from one every ten years to one every three years. The country is now experiencing droughts bi-annually The Zambezi Valley landscape lies in the agroecological regions IV and V, which are characterised by low rainfall and high temperatures. Droughts negatively affect crop and livestock production (WFP, 2014). They also increase the non-availability of water for livestock, human consumption and other domestic uses. Because of limited resources, lack of information and limited technical knowledge, successive droughts weaken smallholder farmers' adaptive capacity (GoZ & UNCT Zimbabwe, 2014). The combined effect of droughts is that they cause changes in crop and livestock prices to levels that may not be affordable for farming households (UNDP, 2016). Droughts cause water scarcity, resulting in communities failing to replenish water in dip tanks. Water scarcity results in livestock farmers failing to dip their cattle and causes the proliferation of pests and diseases (UNDP, 2021). Penultimately, droughts erode savings, livestock and purchasing power, thus adversely affecting coping strategies for farming households (UNDP, 2019b).

3.1.2.2 Floods/Excessive Rains

Floods and droughts mainly characterise climate change in Zimbabwe. The country is experiencing frequent violent storms that culminate in floods (UNDP, 2015). The floods are caused by cyclonic activities that originate in the Mozambican basin. These floods are accompanied by violent storms, hail, and strong winds, which damage crops and cause loss of human lives and property (UNDP, 2016). Most of these floods occur in low-lying areas such as the Zambezi Valley and areas close to major rivers' confluences and downstream of major dams (GoZ & UNCT, 2014). Floods damage crops and human and livestock lives and thus increase the vulnerability of smallholder farmers to further shocks.

3.1.2.3 Crop Diseases and Pests

Zimbabwe is one of the countries in Southern Africa that experiences crop diseases and pests. These pests and crop diseases adversely affect crop production in arid and semi-arid regions such as the Zambezi Valley. In 2017–2018, Zimbabwe experienced the Fall Armyworm, Quelea birds and the Maize stalk borer. Mbire and Kariba experienced medium prevalence, while Binga had a high prevalence of crop and disease infestation (UNDP, 2016). The Armyworm infested maize and sorghum plunging many households into food insecurity. Due to crop and pest infestation, crop yields declined, and food prices become unaffordable for most farming households (UNDP, 2016). As such, crop and pest infestation made farming households unable to adapt to other climate-induced shocks.

3.1.3 Socioeconomic Shocks

3.1.3.1 Sharp Price Increases

Zimbabwe has been experiencing periods of economic and political challenges. These challenges are negatively affecting economic growth. One of the critical results of economic instability is runaway inflation (GoZ & United Nations Country Team -UNCT, 2010). Due to the high inflation, prices of essential goods and crop inputs have been rising at a galloping rate. The government's policy to peg prices for crop outputs like maize while the prices of crop inputs are left to market forces is also driving prices higher as farmers seek alternative markets. Due to sharp price increases, farmers fail to acquire appropriate and adequate units for their agricultural enterprises. Failure by farmers to finance their production results in the contraction of the agricultural sector and migration of crucial labour resources in search of green pastures (GoZ & UNCT, 2014). This erodes farmers' adaptive capacity and safety nets.

3.1.3.2 Livestock Theft

Livestock theft is a shock that rural farming households face mostly during drought years. Droughts cause the decimation of grazing lands and water sources for livestock. As such, livestock roams around the communities to find better grazing lands and water sources (UNDP, 2021). While doing so, the livestock becomes exposed to the risk of theft. The theft of livestock weakens a household's livestock asset base and reduces the farming household's capacity to adapt to climate-induced shocks.

3.1.3.3 Death of a Household Member

The death of a household member weakens a household's capacity to adapt to climate change. A household participating in the ZRBF project may have received a business loan and decides to take part of that money to attend the funeral. This would be done with the hope that the borrower would return the money after the funeral. However, in most cases, the household member who diverted the money to the funeral would face another shock, such as illness or death of another relative, before returning from the funeral. The new shock will cause the project beneficiary to commit less time to the business and enhances his/her chance of defaulting on the loan.

3.1.4 Interventions by Zambezi Valley Alliance and Desired Behavioural Changes

The Zambezi Valley Alliance (ZVA), comprising the Zimbabwe Environmental Law Association (ZELA), Action Aid Zimbabwe, African Breeders Services Total Cattle Management (ABS TCM) and Afrosoft, was implementing the ZRBF program in Binga, Mbire and Kariba districts. Given the erratic rainfall, frequency of shocks and stresses in the Zambezi Valley, the critical interventions that the ZVA were implementing include livestock production, crop production, water sanitation and hygiene, climate-smart agriculture, entrepreneurship and apiculture, productive assets and private sector engagements.

3.1.4.1 Livestock Production

The ZVA consortium supported livestock farmers with interventions meant to improve the breeds and make farmers better adapt to climate change. ZVA facilitated goat and cattle breed improvement for smallholder farming households. The consortium worked with youth, men and women goat groups. Each group of farmers was given a Boer goat or a buck (ZRBF). The group was supported with training on goat husbandry and marketing. The other farmers would bring their she-goats to the farmer with the Boer goat for mating. Through this process, the farmers get improved goats which can fetch higher market prices. ZVA also promoted self-financing for purchasing critical livestock dipping chemicals. The consortium also worked with cattle farmers who were facilitated with access to group loans, cattle pen fattening and linkages to markets. The farmers were also supplied with essential livestock drugs and supply chain linkages through the private sector. For instance, some of the cattle farmers were linked to Montana meats.

3.1.4.2 Crop Production

Farming households were provided with technical advice on crop production. ZVA used the Lead Farmer approach and Crop and Livestock Information Centres (CLICs) to demonstrate technologies to farming households. The CLICs showcased production and processing technologies for small grains (groundnuts, cowpeas, and millet) and fodder. The agronomic advice focussed on conservation agriculture practices and small grains.

3.1.4.3 WASH Interventions

The Zambezi Valley faces a massive challenge with water availability for human and livestock consumption, especially in the hot, dry season. To address this challenge, ZVA partners mobilised communities to construct weir dams to reduce trekking distances to water sources in the dry season. The consortium rehabilitated boreholes for livestock drinking and constructed water troughs at high-yielding boreholes (UNDP, 2019b). Some of the boreholes were fitted with solar panels for pumping water so that they could provide clean water to nearby primary schools.

3.1.4.4 Early Warning System

Binga, Kariba and Mbire districts are very vulnerable to shocks such as floods, droughts and outbreaks of livestock diseases. To strengthen the early warning systems, ZVA provided advisory services to farming households through mobile phone-based platforms (UNDP, 2019b). Farming households were sent bulk short messages warning them of crop and livestock disease outbreaks, rainfall and storms, among others. This assisted the farmers in adapting to the shocks by taking necessary precautions.

3.1.4.5 Private Sector Engagement/Market Linkages

Further to assisting livestock and crop farmers, ZVA linked farmers to crop and livestock markets. The consortia provided technical advice to farmers on how to link with private sector markets. This included negotiation with the private sector and entering contracts. This work linked smallholder millet farmers in Kariba to Delta Beverages. Livestock farmers in the Binga district were linked to Montana Meats, and goat farmers were linked to Zvikomborero Farms (ZRBF, 2019). Other arrangements included the linkage of farmers to suppliers of dipping chemicals such as Lee Coopers.

3.1.4.6 Fodder Production, Preservation, and Feed Preparation

Most livestock farmers rely on natural grazing lands for feeding their livestock. However, these pastures dry up and become scarce during the dry seasons exposing the livestock to the risk of poverty death. The ZVA consortium supported livestock farmers with fodder production, preservation and preparation of livestock feed to minimise the effects of recurring droughts (UNDP, 2019b). The livestock farmers were provided with machines which cut crop residue (stover) into livestock feed and reduced work associated with manual cutting. This intervention helped farmers to supplement their cattle feed during the dry season and protect breeding stock. Fodder production helps to bridge the nutrition gap caused by low and erratic rainfall and enhances farmers' adaptation to climate change. Smallholder farmers can also benefit from selling extra fodder to livestock farmers and generate financial resources which can be used to enhance adaptation (Hove et al., 2003).

3.1.4.7 Climate-Smart Agriculture Practices

ZVA promoted Climate-Smart Agriculture (CSA) practices among smallholder farming households. These interventions focused on crop diversification and improved seeds and small grains. Households that use CSA practices are better able to adapt to climate change.

3.1.4.8 Entrepreneurship and Apiculture

The ZVA consortia organised farmers in Binga, Kariba and Mbire districts into business groups. These were supported with training on business management, leadership and record keeping. The apiculture groups were established in Kariba district to generate income for the farmers and address the challenge of Human Wildlife Conflict. Smallholder farmers who live along wildlife corridors put beehives nearer to their fields to deter elephants. In most cases, the elephants destroy crops just before the harvest and plunge farmers into food insecurity. ZVA facilitated the formation of internal savings and lending schemes (ISALs) for these groups. This was done across the three districts in wards where the project established enterprise groups across various value chains. The purpose of these ISALs was to enable the farmers to access working capital for their groups locally. Formal financial institutions demand collateral, which most farmers cannot provide. As such, ISALs have less stringent lending rules than formal financial institutions.

3.1.5 Components of Adaptive Capacity

Resilience building is premised on three capacities: adaptive, absorptive and transformative. In this section, we focus on adaptive capacity. The section's primary focus is to discuss the components of adaptive capacity and how they affect farming households' capacity to adapt to climate change.

3.1.5.1 Livestock Assets

Livestock assets such as cattle are an essential source of food, income and draft power for smallholder farming households (Oxford Policy Management, 2018). ZVA facilitated the accumulation of livestock assets such as improved goat varieties and chickens to promote income and enhance adaptation to climate change. Before the ZRBF programme, most farming households were not keeping small livestock as a source of income (UNDP, 2020). Keeping small livestock for income enhances a household's income resources from the sale. This raises critical financial resources that can be used to supplement household food requirements (UNDP, 2021). This means households that own livestock can sell the animals to smoothen their consumption during a shock or stress and are therefore better able to adapt to climate change (Nelson et al., 2016). Farmers who own livestock, such as goats and cattle, may dispose of them during shocks and stresses to supplement their household food needs (Zamasiya, Mukamuri & Nyikahadzoi, 2020). If a household is poor, it may be forced by circumstances to sell its livestock to smoothen current consumption, especially to address household food shortfalls during a drought. However, selling of livestock by poor households may spell destitution if the shock persists.

3.1.5.2 Livelihood Diversification

Livelihood diversification is one of the essential strategies for assisting smallholder farmers in adapting to the effects of climate change (Nelson et al., 2016). A diversified household is engaged in more than one livelihood activity or cash source. It provides an effective buffer against unpredictable stresses. The strategies can vary from climate-sensitive to non-climate-sensitive (Béné, Frankenberg & Nelson, 2015). Livelihood strategies are not equally vulnerable to different risks associated with shocks and stresses. A livelihood strategy's effectiveness depends on the household's socioecological context and resource endowments (Nelson et al., 2016). Some contexts may require specialised livelihoods, such as cattle rearing, while others may require a mixture of strategies. Resource-constrained farmers may diversify into other Agri-based livelihoods and remain vulnerable to shocks and stresses. Resource-endowed farmers may venture into non-farm or farm-based livelihood strategies and become less vulnerable. Nelson et al. (2016) observed that crop and livestock-based livelihood diversification in drier areas may not positively affect resource-constrained farmers. We note that livelihood diversification reduces vulnerability when higher returns to livelihood strategies exist.

3.1.5.3 Productive Assets

Productive assets play an important role in building a household's capacity to adapt to climate change (Oxford Policy Management, 2018). Owning productive assets such as ox-drawn ploughs, mobile phones, planters, ridgers, knapsacks, sewing machines, shovels, spades, wheelbarrows, scotch carts and hoes, among others, determines the return a farming household can derive from the land. These implements significantly influence how a household adapts to climate change and recovers from a shock (Guiteras et al., 2015). For instance, ownership of agricultural equipment enables the timely execution of farm operations (UNDP, 2021). Such households can execute specific agricultural processes by substituting human labour and enhancing food security. Likewise, ownership of a mobile phone may assist a household in accessing market information and incur lower transaction costs in marketing and selling of produce (UNDP, 2021). As such, households with productive assets are more likely to be able to adapt to climate change and have a better well-being than households without productive assets.

3.1.5.4 Bridging Social Capital

Bridging social capital concerns connections, relations, or links between individuals (between group connections) who are dissimilar in socioeconomic and other characteristics (Villalonga-Olives et al., 2016). The connections, which may be in the form of friends, relatives and neighbours in a different community, stretch beyond the shared identities and are used for common aims (Guitres et al., 2015). Strong bridging capital creates a safety net upon which farming households can call for help to adapt to climate change (UNDP, 2021). It creates the potential access to crucial resources such as food, information and money outside their constrained environment in times of need for farming households (Putnam, 2000). For instance, friends and neighbours provide critical information which may be difficult to access during a shock. These relationships can be reciprocal in that farming households from another community may also seek help from those they helped in previous seasons. The ability of a farming household to access help from outside one's network is linked to better well-being in times of environmental or social shocks (Villalonga-Olives et al., 2016). Friends and neighbours may provide critical emotional support during a shock. As such, farming households with substantial bridging capital are less likely to use harmful coping strategies, such as food-based or asset-based livelihood coping strategies, to adapt to the impacts of environmental, social and economic shocks and stresses (UNDP, 2021).

3.1.5.5 Linking Social Capital

According to Putnam (2000), linking social capital connects smallholder farmers and outside service providers, such as government officials or NGO workers, who can help them adapt to climate change, if needed. Linking capital provides critical information to farming households that they would otherwise not obtain locally. It helps communities and households to access resources, information and services that are otherwise unavailable through bonding and bridging social capital (Taruvinga et al., 2017). Through linking social capital, farmers can access the information they can use to make informed decisions on their production, which positively affects food security (Aldrich, 2012; UNDP, 2020). Linking capital builds connections between farmers and other actors with different levels of power, for example, public sector (line ministries, research institutes); economic status (private sector agro-processors, marketers, and financial services); and offering different services (civil society players (non-governmental organisations (NGOs), unions and advocacy organisations). Farmers with access to linking capital can better adapt and absorb shocks and stresses than their counterparts (Zamasiya et al., 2021). They are less likely to use harmful coping strategies in the event of a shock or stress to their livelihood system. Despite implementing coping strategies to manage the effects of a shock, linking capital addresses the limitations that farming households face when they adapt to climate change with their means.

3.1.5.6 Human Capital

Human capital depicts a household's level of education in terms of having completed primary or higher education (UNDP, 2021). It determines a household's ability to acquire and use the information to adapt to climate change (Davis et al., 2007; Mulwa et al., 2017). Education improves farmers' access to information, their capacity to make choices, and access to financial and social resources (Zamasiya et al., 2018). It also enhances a farmer's keenness to adopt management practices and technologies that can improve food security in the face of shocks and stresses (Zamasiya et al., 2021). Davis et al. (2007) noted that educated farmers are more nimble, entrepreneurial, and quick to adjust to new realities due to shocks and stresses. Low levels of human capital imply poor and slow response and recovery from shocks and stress. Human capital determines a household's ability to access and utilise higher return activities in the agricultural or non-agricultural sectors. This means that educated farming households are more likely to be better able to adapt to climate change than their counterparts.

3.1.5.7 Access to Financial Services

Smallholder farmers' access to financial services is one of the key drivers of adaptation to climate change. This access can be through community-level institutions that provide credit or savings support. In Zimbabwe, most rural communities are not linked to formal institutions that provide financial services. However, during periods of macroeconomic instability or stress, access to cash or credit from financial institutions plays a vital role in improving adaptation to climate change and the resilience of households (UNDP, 2021). Access to financial resources helps communities to purchase much-needed inputs for their enterprises, sustain enterprises in the event of social and economic disruptions and support livelihood diversification (Gill & Ritchie, 2011). For instance, youth groups running chicken and goat enterprises supported under ZVA were linked to financial services institutions. The linkage to financial services helps the enterprises and the members to better adapt to climate change through access to financial resources. Access to financial resources allows the group members to generate financial resources, which they can use to purchase household food requirements. In the event of a shock, farming households and youth can better cope with the new challenge.

Previous studies have concentrated on understanding determinants of food security by focusing on food consumption scores and Household dietary diversity (Nyikahadzoi et al., 2017; Mango et al., 2014; Shisanya & Mafongoya, 2016; Nyikahadzoi et al., 2017; UNDP, 2021). These studies provide insights into how demographic and socioeconomic variables affect food security and how households adapt to climate change. However, none of these studies provides insight into how farmers' resilience to shocks and stresses can be improved. There is a scarcity of empirical evidence on how adaptive capacities affect food security. The lack of clarity on how components of adaptive capacity contribute to the betterment of household food security indicates that it may be difficult for policymakers and development partners to identify target variables for improving the food security of marginalised communities in similar contexts. This chapter, therefore, seeks to enhance policymakers' and development practitioners' understanding of the determinants of household food security using the coping strategy index. The literature argues that policymakers have a limited understanding of the determinants of household food security for farming households experiencing environmental, economic, and social shocks and stressors (Mabe et al., 2012; Nyikahadzoi et al., 2017). Unless factors determining smallholder farmers' food security are known and addressed, many donor programmes designed to assist farmers in managing shocks and stresses will not benefit the farmers. Therefore, it is critical to determine the drivers of smallholder farmers' food security to shocks and stresses and design strategies to enhance them.

Against this background, this study's main aim was to understand how components of adaptive capacity and project-specific variables condition smallholder farmers' food security. The study used ordinary linear regressions to establish the relationship between the Coping Strategy Index (a proxy for household food security), components of adaptive capacity and project interventions for the Zambezi Valley Alliance Consortia in Binga, Kariba and Mbire districts. Variables hypothesised to affect household food security are presented in Table 3.1. The rest of the chapter is arranged as follows. Section 3.2 discusses the study approach, which includes a description of the site, sampling, and data collection. This is followed by a review of the underlying analytical approaches in Sect. 3.3. Section 3.4 then presents and discusses the results, while conclusions and recommendations follow in Sect. 3.5.

Table 3.1 ZVA summary of CSI across the three OMS rounds

	OMS round		
	OMS1	OMS2	OMS3
Mean CSI score	22.2	22.5	20.9

Source: UNDP, 2019a, b, 2020, 2021 ZRBF Outcome Monitoring Surveys

3.2 Methodology

3.2.1 Description of Study Sites

This study focused on the districts within the Zambezi Valley. Most of the area is under agroecological region V, characterised by an average temperature of 40 °C and rainfall of less than 600 mm per annum. For this chapter, Binga, Kariba and Mbire were selected because they face frequent environmental shocks from droughts and floods. They also have a high potential for livestock production, especially cattle and goats.

Binga, Kariba and Mbire districts lie in the Zambezi Valley in the northern part of Zimbabwe. Mbire District is in Mashonaland Central Province at an altitude of 400 metres above sea level and covers an area of approximately 4700 km^2 (Gumindoga et al., 2020). The district lies in agroecological region V, with highly variable and erratic rainfall, which ranges between 550–850 mm per year. The district is prone to seasonal floods and droughts, making rain-fed crop production risky. The predominant soils are calcic solonetz, which is acidic with high nutrient depletion and is easily erodible (Musiwa & Mhlanga, 2020). The primary livelihood activities in the districts are livestock and crop production. Binga is a semi-arid district in the Zambezi Valley. The district is dominated by Tonga-speaking people who rely on fishing and riverine agriculture for survival. The impoverished soils and the erratic rainfall make rainfed agricultural production almost impossible. The primary land use is wildlife conservation (Dzingirai, 2003). Kariba district lies in the north-eastern part of the Zambezi Valley alley area. The district has three communal areas: Omay, Kanyati and Gatshe and an urban area. Kariba is generally a semi-arid district with very high temperatures, high rainfall variability of <650 mm per year and poor soils. The farmers practice livestock production and crop production.

3.2.2 Data Sources

This study used quantitative data from the Zimbabwe Resilience Building Fund (ZRBF) programming's Outcome Monitoring Survey 3 (OMS3). ZRBF engaged enumerators from Zimbabwe Vulnerability Assessment Committee (Zim VAC), ZRBF partners and Monitoring and Evaluation staff to collect data during fieldwork. TANGO International and ZRBF Project Management Units provided technical support and oversight. The study is based on Outcome Monitoring Survey (OMS) data for 2019, 2020 and 2021. The OMS used a structured survey questionnaire consistent with the ZRBF indicators. The surveys collected data on beneficiaries' demographics, human capital, project interventions that focus on adaptative, absorptive and transformative capacity, ownership of productive assets, social capital (bridging, bonding, and linking), food security, coping strategies, and exposure to shocks. The data was collected through face-to-face interviews with an electronic questionnaire form programmed using Open Data Kit (ODK). The study used mobile devices and electronic questionnaires for data collection, allowing data validation rules and consistency checks to be implemented and minimising data collection errors. Data were uploaded to the server maintained by the United Nations Development Programme (UNDP) to ensure data protection and confidentiality. The sample sizes for ZVA that were reached in the three outcome monitoring surveys are 485 in OMS1, 449 in OMS2 and 478 in OMS3 (UNDP, 2021).

3.2.3 Analytical Framework

The study used the ordinary least squares (OLS) regression (Greene, 2018) to investigate the drivers of household food security for smallholder farmers facing climate-induced shocks and stresses. In this chapter, the OLS regression was used for each OMS round to investigate how components of adaptive capacity and project interventions affected food insecurity as measured by the Coping Strategy Index. The use of the OLS model allows the researcher to understand how the independent variables behaved across the three Outcome Monitoring Surveys. This is because, the study site experienced unique challenges in each OMS namely, a drought, COVID-19 lockdown and socioeconomic crisis (UNDP, 2021). While a pooled OLS regression analysis would be proper to take advantage of the panel data set, the dataset from only one consortium is insufficient to run the robust pooled regression model. Under such circumstances, an OLS model becomes handy. The strength of the OLS regression is that it can be easily used to model a linear relationship between a continuous dependent variable and one or more independent variables (Greene, 2018).

The OLS model can be expressed as follows:

Y* = f(component of adaptive capacity, project level variables)
Let $Y^* = BX_i + E_i$ general model

Therefore, the general model can be rewritten as follows:

$$\text{Food Security} = b_0 + b_1 x_1 + b_2 x_2 + \cdots . b_n X_n + e_t \tag{3.1}$$

$$CSI = b_0 + b_1 x_1 + b_2 x_2 + \cdots . b_n X_n + e_t \tag{3.2}$$

where Y^* = food security, (CSI) = dependent variable

$X_{1.....Xn}$ parameters of independent variables to be estimated
b_0 = intercept
b_1...bn coefficients for independent variables
b = vector of coefficients to be estimated
e_i is the error term

3.2.4 *Dependent Variables*

3.2.4.1 Coping Strategy Index

The Coping Strategy Index (CSI) is a food security score that assesses the extent to which households may use negative or harmful coping strategies when they do not have enough food or enough money to buy food (Maxwell & Caldwell, 2008). The CSI measures the frequency and severity of coping behaviours. A high CSI means poor food insecurity, and a low CSI suggests improved food security.

According to Table 3.1, the CSI score for ZVA was 22.2 in OMS1, 22.5 in OMS2 and declined to 20.9 in OMS3. The results generally depict a decline in CSI among farming households. This suggests that there was a general improvement in food security. The results indicate a decline in negative coping strategies to address food insecurity.

3.3 Independent Variables

The independent variables used in this study were chosen based on the literature review and the project context (Table 3.2).

Table 3.2 Description of dependent and independent variables

Dependent variables	Description	Expected sign
Outcome variables	*Food consumption score* *Coping strategy index*	
Independent variables		
Components of adaptive capacity	Linking social capital	+
	Bridging capital	+
	Exposure to information	+
	Livelihood diversification	+/–
	Access to financial services	+/–
	Human capital	+
	Livestock assets	+
	Productive assets	+/–
Project level variables	*Climate-smart agriculture practices*	+/–
	Goat improvement	+
	Participate in fodder production	+
	Participate in fodder preservation	+
	Participate in goat breed improvement	+

3.3.1 Data Analysis

Data analysis was conducted in STATA version 15. Before conducting data analysis, correlation analysis was conducted among independent variables to address the problem of multi-collinearity. A cut-off point/threshold of 0.7 was used to drop the undesirable/variables with a high correlation. There were no variables that were dropped based on this criterion. In terms of project interventions, any interventions that were being practised by less than 10% of the respondents were dropped in the analysis. Using this approach, artificial insemination was dropped from the project variables. Unitary analysis was also conducted to determine the means and proportions of the independent variables. Finally, several OLS regression equations were run to identify the adaptive capacity components and project-specific variables that condition household food security. This process was done for OMS1, OMS2 and OM3. The reason for analysing the three OMS rounds was to understand how components of adaptive capacity affect the coping strategy index for marginalised communities. The process is also crucial for identifying how the components of adaptive capacity condition household adaptation to climate change-induced shocks and stresses.

3.4 Results and Discussion

3.4.1 Descriptive Analysis of the Independent Variables

Table 3.3 presents the descriptive statistics for the independent variables. The results show that from rounds 1 to 3, adaptive capacity increased from 24.7 to 30.7. The increase was statistically significant. Linking capital increased from 18.2 to 31.1. Bridging capital increased from 28.1 to 33.0. Exposure to information improved from 22.5 to. 44.8. Livelihood diversification increased from 8.6 to 8.8. Access to financial services fell from 20.15 to 201. Human capital increased from 51.4 to 52.7. The results showed that most adaptive capacity components increased statistically significantly from round 1 to round 3.

3.4.2 Model Estimates for the Determinants of the Coping Strategy Index

Table 3.4 presents the linear regression results on the CSI determinants for each OMS round. The results showed that in OMS1, bridging capital ($p < 5\%$) and participation in goat improvement ($p < 5\%$) had negative and statistically significant relationships with food insecurity. In OMS2, access to financial services ($p < 10\%$) had a negative and statistically marginally significant relationship with food

Table 3.3 Descriptive statistic

Independent variables		OMS1	OMS2	OMS3
		Mean/prop	Mean/prop	Mean/prop
	Adaptive capacity	24.7	27.0	30.7**
	Linking capital	18.2	28.4**	31.9
	Bonding capital	28.1	28.0	33.0
	Exposure to information	22.5	39.7***	44.8
	Livelihood diversification	8.6	7.0**	8.8
	Access to financial services	20.5	15.6	20.1
	Human capital	51.4	54.5*	52.7
Project level variables	Participate in goat breeding	14.84	20.27	26.99
	Practice fodder production	39.47	23.71	25.94
	Practice fodder preservation	43.84	46.75	24.06
	Climate-smart agricultural practices	2.58	3.89	4.59

insecurity. The results for OMS3 showed that human capital ($p < 5\%$), exposure to information ($p < 5\%$), productive assets ($p < 1\%$), linking capital ($p < 10\%$) and livestock assets ($p < 10\%$) had a negative and statistically significant relationship with CSI. However, only climate-smart agricultural practices ($p < 1\%$) had a positive and statistically significant relationship with CSI.

3.4.3 Bridging Capital

Study findings showed that bridging social capital has a negative and statistically significant relationship ($p < 5\%$) with CSI. These results are consistent with the findings of UNDP (2020) and UNDP (2021). These results mean that farming households with strong bridging capital are less likely to use food-based and asset-based livelihood coping strategies. Such households are better able to adapt to climate change than their counterparts. This is because they can rely on the linkages with relatives and non-relatives from their communities and outside their communities whom they can ask for help to adapt to climate change. As such, farming households with strong bridging capital are more likely to be resilient to shocks and stresses than their counterparts.

3.4.4 Goat Improvement

The results showed that goat improvement has a negative and statistically significant relationship ($p < 5\%$) with CSI. The results corroborate with the findings of UNDP (2021). These results imply that farming households that participated in goat

Table 3.4 OLS regression results for the determinants of CSI

	OMS1	OMS2	OMS3
Independent variables	Coef.	Coef.	Coef.
Access to financial services	−3.9442 (6.0236)	−13.4472* (6.8613)	−2.8684 (3.7487)
Human capital	−1.1223 (2.3742)	−0.8969 (1.9867)	−1.9363** (0.7967)
Livelihood diversification	−0.3707 (0.7309)	0.5059 (0.5887)	−0.1662 (0.2239)
Livestock assets	−0.0520 (0.3459)	−0.2135 (0.2581)	−0.2376* (0.1370)
Linking capital	1.6637 (1.3812)	−0.1968 (0.9520)	−0.7342* (0.4194)
Bridging capital	−5.0890** (2.0158)	0.9571 (1.6596)	−0.3904 (0.6207)
Goat improvement	−10.9473** (5.1709)	−2.9453 (3.3946)	1.6130 (2.3268)
Participate in fodder production	3.1594 (7.2950)	2.5056 (5.4772)	−2.3910 (2.1976)
Participate in fodder preservation	6.5722 (7.7470)	8.8111 (5.5832)	−2.0867 (2.5527)
Climate-smart adaptations	1.4133 (0.9802)	−0.8125 (0.8779)	2.1374*** (0.3570)
Productive assets	−1.1368 (1.0756)	−0.8339 (0.6937)	−1.3464*** (0.3322)
Exposure to information	−0.7662 (0.6339)	0.4758 (0.5089)	−0.5083** (0.2395)
Constant	33.7692 (12.3814)	23.3385 (7.3173)	32.6552 (2.7640)

Figures in parentheses are standard errors
*** 1% level of significance; ** 5% level of significance; * 10% level of significance
OMS1: $F(12; 48) = 1{,}75$; Prob $>F = 0.0850$; $R^2 = 0.3045$
OMS2: $F(12; 47) = 1{,}70$; Prob $>F = 0.0963$; $R^2 = 0.3032$
OMS3: $F(12; 465) = 8.35$; Prob $>F = 0.0.0000$; $R^2 = 0.17.72$

breed improvement are more likely to better adapt to climate change than their counterparts. This is because when households participate in goat breed improvement, the offsprings fetch higher prices than their local ones. The practice of farmers participating in goat breed improvement opens new income opportunities for the household through selling improved breeds of offsprings. For instance, a farming household can do stress disposal of goats to raise financial resources for basic household needs. Therefore, a farming household can better adapt during a shock as they have more income resources. The farming household may use the additional financial resources from goats to purchase household food requirements (Zamasiya et al., 2021). Using financial resources from stress disposal of improved goats limits the use of negative coping strategies by farming households and improves their resilience to shocks.

3.4.5 Access to Financial Resources

The study findings showed that access to financial resources has a negative and marginally statistically significant relationship (p < 10%) with CSI. These results are consistent with the findings of UNDP (2020). These results mean that farming households who could access financial services such as savings or loans were able to adapt to climate change better than their counterparts. This is because access to savings helps farming households preserve their savings in foreign currency in volatile macroeconomic conditions (UNDP, 2021). The households can generate more income by providing loans in ISAL groups. Access to loans also improves a household's income and expenditure. They can then use the increased income to adapt to climate change. Such households are less likely to use negative coping strategies. Therefore, households with access to savings are more resilient to shocks than their counterparts. However, it should be noted that in the Zambezi Valley, it was challenging for households to access loans from formal financial institutions due to the long distances to the financial institutions (UNDP, 2021).

3.4.6 Human Capital

Results showed that goat improvement has a negative and statistically significant relationship ($p < 5\%$) with CSI. The results corroborate the findings of UNDP (2021), which noted that human capital plays a pivotal role in adaptation to climate change. These results imply that farming households who are more learned/have accumulated human capital through attending workshops, crop and livestock demonstrations at CLICs and exchange visits are more likely to understand livestock disease calendars. They can then use their accumulated human capital to make informed decisions on how to adapt to climate change. For instance, human capital helps a goat farmer to understand goat disease calendars. As such, farming households with high levels of human capital can understand the impact of a disease outbreak and its repercussions on the household's livelihoods. As such, the ZVA interventions facilitated farming households to make informed decisions on their livelihood activities better and adapt to climate change.

3.4.7 Productive Assets

The study findings show that productive assets have a negative and statically statistically significant relationship ($p < 1\%$) with CSI. These results are consistent with the findings of UNDP (2020) and suggest that at the time of OMS3, farming households under the ZVA had accumulated productive assets. The accumulation of productive assets was making them better prepared to manage shocks. For instance, ownership of a mobile phone may help a household to improve access to early warning information and decide on the time of planting (UNDP, 2021). Ownership of an ox-drawn plough may mean that such farming households can take advantage of rainfall and capitalise on that. Farming households who do not own farming equipment are less able to adapt to climate change, affecting their household well-being. Poorer households may resort to stress selling productive assets to raise financial resources to purchase food (UNDP, 2020). This may weaken their adaptive capacity in future shocks.

3.4.8 Exposure to Information

The study findings showed that exposure to information has a negative and marginally statistically significant relationship ($p < 5\%$) with CSI. These results are consistent with the findings of UNDP (2021). These results mean that at the time of OMS3, farming households were exposed to more information under the ZRBF programme. This information may include rainfall patterns, disease outbreaks and early warning. Farming households with higher exposure to information are better placed to use it to make decisions on their livelihood strategies. For instance, a farming household that received early warning information may use it to decide on the climate-smart adaptation practice they will use for the coming season. This makes them better adapt to climate change than their counterparts. Exposure to information also enhances the uptake of programmes such as the free inputs under the "Pfumvudza" Programme. As such, households with higher exposure to information are more likely to be resilient to climate change than their counterparts.

3.4.9 Livestock Assets

Results showed that livestock assets have a negative and statistically significant relationship ($p < 10\%$) with CSI. The results corroborate the findings of UNDP (2021), which note that small livestock such as goats and chickens play a vital role in raising household income for basic needs. ZVA facilitated the accumulation of small livestock assets such as improved goat breeds and chickens. Through the ZVA interventions, farming households in the Zambezi Valley have now transformed into livestock-rearing for businesses (UNDP, 2021). These households can now raise financial resources, which they can use to adapt to climate change. Households with additional income resources from selling goats and chickens are less likely to use negative coping strategies in response to shocks because of the ZRBF interventions. Livestock, such as cattle, provides draft power for field preparation for planting (UNDP, 2021). This implies that farming households that own cattle can better take advantage of moisture opportunities and plant crops. By so doing, these households are better able to adapt to climate change than their counterparts.

3.4.10 *Linking Social Capital*

Research findings showed a negative and statistically significant relationship ($p < 10\%$) between linking capital and CSI. These results mean that households with access to linking capital were more food secure than their counterparts. The probable explanation for this result is that the ZVA interventions strengthened linkages between farming households and institutional stakeholders through capacity building and programme interventions. The interventions by ZVA improved relations between farmers, extension officers and other support services. As such, there was an improvement in service provision to the farming households (UNDP, 2020). Moreover, farming households with access to linking social capital are better able to adapt to climate change through support from institutional stakeholders. Under these circumstances, they are less likely to use negative coping strategies.

3.5 Conclusion and Recommendations

In this chapter, the study used the ordinary least squares regression approach to identify the role of components of adaptive capacity and critical project interventions by the Zambezi valley Alliance under the Zimbabwe Resilience Building Fund programme on food security. The study used data from the three OMS rounds. The study findings showed that ownership of productive assets, exposure to information, human capital, bridging capital and goat improvement are crucial to enhancing a household's adaptation to climate change. The study concludes that the effect of a component of adaptive capacity is conditional to the type of shock that a household is exposed to; for instance, during periods of economic stability such as OMS1, goat improvement and bridging capital played an important role in enhancing adaptation to climate change. However, during social stress (OMS2), such as COVID-19, where it is difficult to conduct business and travel across districts, access to financial resources was key to enhancing adaptation to climate change. During currency instability and other socioeconomic challenges (OMS3), investments in human capital, livestock assets, linking social capital and productive assets were crucial to enhancing adaptation to climate change. Based on these findings, we, therefore, recommend the following.

3.5.1 *Promote Integrated Small Livestock Asset Accumulation*

This research has stressed the importance of livestock asset accumulation, goat breed improvement, linking capital, exposure to information, access to financial services and human capital and the pivotal role of improved goats' intervention in assisting farmers in adapting to climate change. To build the capacity of farmers to adapt to climate change and improve their well-being, there is a need for deliberate efforts to promote an integrated livestock asset accumulation model. This model can ride on critical interventions or activities that should be layered with the goat improvement intervention. In the first stage, potential recruits should be exposed to information about the role of small livestock accumulation in enhancing adaptation to climate. In the second stage, the development partner should then facilitate the establishment of farmers' groups. This process should be conducted with the help of local service providers such as Agritex and Livestock Production. Smaller groups of farmers can be assisted to access improved buck breeds such as the Matabele, Boer and Namib. As a sustainability measure, the farmers should be requested to contribute at least 30%, and the development partner should meet 70% of the cost of acquiring the goats. This model will ensure that the farmers commit themselves to the project. The model has worked with other development partners on livestock pass-on schemes (Zamasiya et al., 2018).

Once the farmers are organised into small livestock groups, the development partner should focus on exposure to information and human capital accumulation. This can be implemented through the establishment of livestock learning centres. These should be the primary demonstration sites/innovation hubs where the groups of farmers can learn about the small livestock improvement model. The learning centre approach enables the exposure of farming households to information. Successful adaptation to climate change requires that farming households get exposed to information on various topics. This information is critical for the success of the livestock accumulation model. The human capital accumulation component can be addressed through training on livestock management, marketing, business management, exchange visits and contracts. The last component of the integrated model should then be to speak of issues of access to financial resources for the small livestock groups. Given the challenges with access to formal financial institutions in smallholder farming communities, there is a need to build the capacity of the smallholder livestock to organise and run ISALs. The development partner may assist the ISAL with a revolving stress fund. This fund can enhance the group's access to financial services to purchase she-goats, livestock chemicals and meet other running costs. Through this model, adaptation to climate change by smallholder farming households can be addressed.

References

Aldrich, D. P. (2012). *Building resilience: Social capital in post-disaster recovery*. University of Chicago Press.

Becchetti, L., & Castriota, S. (2011). Does microfinance work as a recovery tool after disasters? Evidence from the 2004 tsunami. *World Development, 39*(6), 898–912.

Béné, C., Frankenberger, T., Nelson, S. (2015). Design, monitoring and evaluation of resilience interventions: Conceptual and empirical considerations. *The Institute of Development Studies and Partner Organisations. Report.* https://hdl.handle.net/20.500.12413/6556

Bene, C., Wood, G. R., Newsham, A., & Davies, M. (2012). *Resilience: New utopia or new tyranny? Reflection about the potentials and limits of the concept of resilience in relation to vulnerability reduction programmes*. Institute of Development Studies.

Davis, B., Winters, P., Carletto, G., Covarrubias, K., Quinones, E., Zezza, A., Stamoulis, K., Bonomi, G., & DiGiuseppe, S. (2007). Rural income-generating activities: A cross country comparison. In *Background paper for the world development report 2008*.

Dzingirai, V. (2003). CAMPFIRE is not for Ndebele Migrants': The impact of excluding outsiders from CAMPFIRE in the Zambezi Valley. *Journal of Southern African Studies, 29*(2), 445–459.

FAO. (2015). *Regional overview of food insecurity: African food security prospects brighter than ever*. Food & Agriculture Organisation.

Frankenberger, T., Mueller, M., Spangler, T., & Alexander, S. (2013). Community resilience: conceptual framework and measurement feed the future learning agenda. Rockville, MD: Westat.: USAID.

Gill, D. A., & Ritchie, L. A. (2011). *Considering community capitals in disaster recovery and resilience*. Public Entity Risk Institute (PERI).

Government of Zimbabwe and UNCT Zimbabwe. (2010). *Country analysis report for Zimbabwe*. Harare.

Guiteras, R. P., Jina, A. S., & Mobarak, A. M. (2015). Satellites, self-reports, and submersion: Exposure to floods in Bangladesh. *American Economic Review: Papers and Proceedings, 105*, 232–236.

GoZ & UNCT. (2014). *Government of Zimbabwe*. Zimbabwe Country Analysis: Working Document.

Gumindoga, W., Murwira, A., Rwasoka, D. T., Jahure, F. B., & Chikwiramakomo, L. (2020). The Spatio-temporal soil moisture variation along the major tributaries of Zambezi River in the Mbire District, Zimbabwe. *Journal of Hydrology: Regional Studies, 32*, 100752–100776.

Greene, W. (2018). *Econometric analysis* (8th ed.). Pearson Education Limited.

Hove, L., Franzel, S., & Moyo, P. S. (2003). Farmer experiences in the production and utilisation of fodder trees in Zimbabwe: Constraints and opportunities for increased adoption. *Tropical Grasslands, 37*(279–283), 279.

IPCC. (2014). *Climate change 2014 synthesis report. Summary for policy makers*. Intergovernmental Panel on Climate Change.

Koomson, I., Villano, R. A., & Hadley, D. (2020). Effect of financial inclusion on poverty and vulnerability to poverty: Evidence using a multi-dimensional measure of financial inclusion. In *Network for Socioeconomic Research and Advancement Working Paper, NESRA/WP/20/001* (pp. 1–25).

Mabe, F., Sarpong, D., & Osei-Asare, Y. (2012). Adaptive capacities of farmers to climate change adaption strategies and their effects on rice production in the northern region of Ghana. *Russian Journal of Agriculture Socio-Economics Sciences, 11*(11), 9–17.

Mango, N., Zamasiya, B., Makate, C., Nyikahadzoi, K., & Siziba, S. (2014). Factors influencing household food security among smallholder farmers in the Mudzi district of Zimbabwe. *Development Southern Africa, 31*(4), 625–640. https://doi.org/10.1080/0376835X.2014.911694

Maxwell, D., & Caldwell, R. (2008). *Coping strategies index: Field methods manual* (2nd ed.) Retrieved from http://files.ennonline

Musiwa, A. R., & Mhlanga, W. (2020). Human-wildlife conflict in Mhokwe Ward, Mbire District, north-East Zimbabwe. *African Journal of Ecology, 00*, 1–10. https://doi.org/10.1111/aje.12774

Mulwa, C., Marenya, P., Bahadur, D., & Kassie, M. (2017). Climate Risk Management Response to climate risks among smallholder farmers in Malawi: A multivariate probit assessment of the role of information, household demographics and farm characteristi. *Climate Risk Management, 16*, 208–221.

Nelson, S., Frankenberger, T., Langworthy, M., Finan, T., & Bower, T. (2016). The effect of livelihood diversity on recovery and shock impact in Ethiopia, Kenya and Uganda. Report prepared by the technical consortium, a project of the CGIAR. In *Technical report series no 2: Strengthening the evidence base for resilience in the horn of Africa*. A joint International Livestock Research Institute (ILRI) and TANGO International Publication.

Nyikahadzoi, K., Adekunle, A., Fatunbi, O., & Zamasiya, B. (2017). Promoting production and marketing of root crops in southern Africa in a changing climate using integrated agricultural research for development (iar4d) pathway. *African Journal of Food, Agriculture, Nutrition and Development, 17*(1), 11787–11802. https://www.ajol.info/index.php/ajfand/article/view/153805

Oxford Policy Management Limited. (2018). *Zimbabwe resilience building fund impact evaluation baseline report.*

Sam, A. S., Abbas, A., Padmaja, S. S., Kaechele, H., Kumar, R., & Müller, K. (2018). Linking food security with Household's adaptive capacity and drought risk: Implications for sustainable rural development. *Social Indicators Research.* https://doi.org/10.1007/s11205-018-1925-0

Shiferaw, B., Tesfaye, K., Kassie, M., Abate, T., Prasanna, B. M., & Menkir, A. (2012). Managing vulnerability to drought and enhancing livelihood resilience in sub-Saharan Africa: Technological, institutional and policy options. *Weather and Climate Extremes, 3*, 67–79.

Shisanya, S., & Mafongoya, P. (2016). Adaptation to climate change and the impacts on household food security among rural farmers in uMzinyathi district of Kwazulu-Natal, South Africa. *Food Security, 8*, 597–608.

Taruvinga, B., Ndou, P., Hlerema, I., Maraganedzha, T., Du Plooy, P., & Venter, S. (2017). Fostering linking social capital for successful agricultural development projects in South Africa. *Agrekon, 56*(1), 28–39.

UNDP. (2015). *Zimbabwe resilience strategic framework.* UNDP.

UNDP. (2016). Mapping of selected hazards affecting rural livelihoods in Zimbabwe. In *A district and ward analysis.* UNDP.

UNDP. (2019a). ZRBF resilience knowledge hub. In M. Langworthy, T. Aziz, & J. Stack (Eds.), *ZRBF outcome monitoring survey: Round one program learning report.*

UNDP. (2019b). ZRBF resilience knowledge hub. In N. Mock, J. Stack, & A. Sundsmo (Eds.), *Assessment of the Zimbabwe resilience building fund crisis modifier mechanism.*

UNDP. (2020). ZRBF resilience knowledge hub. In M. Langworthy, K. Fox, S. Martin, L. Woodson, T. Aziz, R. Alhaddad, E. Cuellar, C. Hein, & J. Stack. *ZRBF outcome monitoring survey: Round two program learning report.*

UNDP. (2021). ZRBF resilience knowledge hub. In S. Martin, J. Stack, S. Nelson, & M. Langworthy (Eds.), *ZRBF outcome monitoring survey: Round three program learning report.*

Villalonga-Olives, E., Adams, I., & Kawachi, I. (2016). The development of a bridging social capital questionnaire for use in population health research. *SSM – Population Health, 2*, 613–622. https://doi.org/10.1016/j.ssmph.2016.08.008

WFP. (2014). *Zimbabwe: Results of exploratory food and nutrition security analysis.* WFP.

Zamasiya, B., Nyikahadzoi, K., & Mukamuri, B. (2018). Drivers of adaptation to climate change by smallholder farmers in smallholder farming communities: A case study of Hwedza district in Zimbabwe. *Zambezia Journal of Humanities, 45*(2), 75–92.

Zamasiya, B., Nyikahadzoi, K., & Mukamuri, B. B. (2020). Drivers of Level of Adaptation to Climate Change in Smallholder Farming Systems in Southern Africa: A Multi-Level Modelling Approach. *African Handbook of Climate Change Adaptation*, 1–12.

Zamasiya, B., Nyikahadzoi, K., & Mukamuri, B. B. (2021). Determinants of household food security using a hierarchical linear modelling technique: A case study of smallholder farmers in Hwedza district in Zimbabwe. In K. Nyikahadzoi & B. B. Mukamuri (Eds.), *Building climate change resilient communities in Africa: Insights from Zimbabwe's urban and rural areas.* Konrad Adenauer Stiftung.

Chapter 4
Water-Based Infrastructure and Resilience in Selected Zimbabwe Resilience Building Fund (ZRBF) Consortiums in South-Western Zimbabwe

Sandra Bhatasara and Praise Karuma

4.1 Introduction

Water is essential for socioeconomic advancement, food security and the maintenance of healthy ecosystems. It is crucial for alleviating the global disease load and enhancing health, welfare and productivity of populations (United Nations Water, 2019). Climate change influences ecosystems, human communities and economies in several ways, with water serving as the principal medium for these effects (Caretta et al., 2022). Climate change influences and is influenced by global water resources, diminishing the predictability of water availability and impacting water quality (UNESCO-UNWater, 2020). Climate change exacerbates the frequency of extreme weather events, jeopardising sustainable socio-economic development and global biodiversity, with significant repercussions for water resources. Approximately 50% of the global population, totalling 8 billion individuals, is projected to encounter significant water scarcity for a portion of the year as a result of climatic and non-climatic influences (Caretta et al., 2022). Climate change is amplifying variability in the hydrological cycle, thereby precipitating extreme weather events, diminishing the predictability of water availability, degrading water quality and jeopardising sustainable development, biodiversity and the enjoyment of human rights to safe drinking water and sanitation globally (UN Water, 2019). Urama and Ozor (2010) emphasised that climate change is exerting several immediate and long-term effects on water resources in African nations. The impacts disproportionately affect impoverished and vulnerable communities, exacerbated by factors such as population growth, unmanaged migration, alterations in land use, diminished soil health,

S. Bhatasara (✉)
Marondera University of Agricultural Sciences and Technology, Marondera, Zimbabwe

P. Karuma
Department of Community and Social Development, University of Zimbabwe, Harare, Zimbabwe

L. Mhlanga et al. (eds.), *Building Resilience to Climate Change in Selected Districts in Zimbabwe*, https://doi.org/10.1007/978-3-031-80529-5_4

intensified groundwater extraction, extensive ecological degradation and biodiversity loss, which affects the lives and the livelihoods of the people (UN Water, 2019).

There is growing focus on adaption technology within the water sector. The signatories of the United Nations Framework Convention on Climate Change (UNFCCC) see water as a critical focus for intervention, with 119 Parties designating water as a priority under the adaptation segment of their Intended Nationally Determined Contributions (INDC). Enhancing the capacity to adapt to climate changes through suitable technology is essential for sustainable development and the survival of global societies (UN Water, 2019). Access to a varied range of suitable and cost-effective adaptation technologies and practices across different contexts is essential (Caretta et al., 2022). This chapter examines water-based infrastructure and the many water-related resilience capacities that contribute to resilience outcomes. ZRBF has significantly invested in enhancing water infrastructure accessibility. This chapter addresses the following question: What water-related infrastructure measures have enhanced the capacity of households and communities to withstand or adapt to shocks and stresses? (ii) In what manner are the water initiatives enhancing resilience capacities? and (ii) what incentivises farmers to participate in their maintenance? The emphasis on resilience is guided by the enquiries in the qualitative instruments (see to ZRBF OMS3 report). This chapter aims to identify water-related stresses and shocks as well as various water-based methods; ascertain the motivations for endorsing different water technologies; analyse resilience outcomes; and recognise hurdles to resilience.

4.2 Background to ZRBF Programming

The Zimbabwe Resilience Building Fund (ZRBF) was established in response to the significant economic, environmental and social shocks impacting rural communities in Zimbabwe (ZRBF, 2021a). The fund is administered by the United Nations Development Programme (UNDP) and financed by the Foreign Commonwealth and Development Office (FCDO), the European Union (EU), the Embassy of Sweden, the Government of Denmark and the United Nations Development Programme (UNDP). The ZRBF is executed in partnership with the Government of Zimbabwe, under the leadership of the Ministry of Lands, Agriculture, Fisheries, Water, Climate, and Rural Resettlement (MLAFWCRR). The Zimbabwe Resilience Building Fund is a long-term development project aimed at enhancing the capacity of communities to safeguard development achievements against recurring shocks and strains, hence facilitating their contribution to Zimbabwe's economic development. This is aimed to be achieved through the collaborative execution of three interconnected multi-sectoral outputs involving several stakeholders. The objective was intended to be reached through multi-stakeholder implementation of three interlinked multi-sectorial outputs, namely:

(i) Application of evidence in policymaking for resilience increased.
(ii) Absorptive, adaptive and transformative capacities of at-risk communities increased and improved.

(iii) Timely and cost-effective response to emergencies rolled out via existing safety net and other relevant programmes.

ZRBF has coordinated initiatives centred on three interconnected elements: establishing evidence, enhancing long-term household and community resilience and implementing a crisis modifier system that offers agile and adaptable funding during periods of shocks and stressors. ZRBF facilitated the execution of resilience-building initiatives throughout 18 rural districts in Zimbabwe through 7 project consortia. The consortia include those that are listed below as follows:

- ECRAS: Enhancing Community Resilience and Sustainability
- MELANA: Matabeleland Enhanced Livelihoods, Agriculture and Nutrition Adaptation
- ZVA: Zambezi Valley Alliance for Building Community
- BRACT: Building Resilience through Improving the Absorptive and Adaptive Capacity for Transformation
- ECRIMS: Enhancing Community Resilience and Inclusive Market Systems
- SIZIMELE: Sizimele Action for Building Resilience in Zimbabwe
- PROGRESS: Programme for Growth and Resilience

The Fund is also facilitating national surveys essential for resilience programmes, including livelihoods and vulnerability assessments, poverty surveys and agriculture-related surveys. This chapter is derived from findings of a mixed-method study on resilience enhancement initiatives funded by the ZRBF, utilising three rounds of a home panel survey and two rounds of qualitative data collection. Three Outcome Monitoring Surveys (OMS) were employed to gather household data for annual reporting on 11 primary indicators and to assess the impact of ZRBF-funded initiatives on those variables. The results from the initial round of the OMS, conducted in April 2019, underscored the necessity for qualitative research to comprehensively comprehend the factors and influences of transformation. A qualitative element was used in the Outcome Monitoring Rounds Surveys 2 and 3 that were carried out across Zimbabwe.

4.3 Literature Review

4.3.1 Water-Based Adaptation to Climate Change in Africa

Climate change affects global water supplies in intricate manners, primarily influencing the timing, quantity, form and intensity of precipitation (Fecht, 2019). Climate change–induced alterations typically exert pressure on drinking water resources, agricultural output, property valuations and several other facets of human existence. Recent studies have demonstrated that significant regional water resources are susceptible to alterations in temperature and climatic patterns (Urama & Ozor, 2010). Caretta et al. (2022) assert that "climate change is acknowledged as one of the most significant concerns anticipated to impact hydrological processes, including precipitation and evaporation." Water is essential for the proper operation of ecosystems and human socioeconomic activities (United Nations Water, 2019). It serves as a resource for agriculture and food production, energy generation, residential applications and industrial manufacturing. Consequently, water shortages have devastating effects on the livelihoods of individuals, particularly in low-income regions globally (ZRBF, 2021a, b, c). Hajkowicz and Higgin (2008) elucidate that water shortage in rural regions has significantly affected individuals' lives and livelihoods, as the majority of livelihood options in these areas are reliant on water and primarily depend on it for production activities. Traep and Bakkegaard (2015) assert that water is a crucial element for production and domestic survival in numerous places globally.

Greenhouse gas emissions into the atmosphere have significantly impacted the water cycle, leading to a reduction in the availability of fresh water globally (Xia et al., 2016). All emissions scenarios forecast a reduction in the availability of renewable surface and groundwater resources for a substantial portion of the world population (ZRBF, 2021a, b, c). Every increment of global warming is anticipated to reduce the availability of renewable water resources by a minimum of 20%, affecting an additional 7% of the global population (IPCC, 2021). The ZRBF OMS 3 study (2021b) identified the alteration of rainfall patterns due to climate change as a contributing factor necessitating water adaptation in Southern Africa. Alterations in rainfall unpredictability and elevated temperatures due to climate change are anticipated to lead to a heightened incidence of water scarcity and droughts (UNFCCC, 2014). The threat of water shortage is intensified by socioeconomic factors like population growth, economic development, increased per capita water consumption and the expansion of agricultural and industrial activities (IPCC, 2014). The ZRBF OMS 2 report (2020) indicates that due to the scarcity of water resources in many rural areas of Southern Africa, there is a necessity for the implementation of definitive water adaptation technologies to address the challenges posed by changing climatic conditions on the region's rural economies, lives and livelihood options.

Most of the projects that are run by ZRBF in Zimbabwe are largely water dependent and have been affected by the scarcity of water in the region due to climate change. As noted in the ZRBF OMS 3 report (2021c), cattle, goats and poultry are raised in all ZRBF intervention areas, which clearly shows how livestock is a major asset for rural households, but of late these have been affected by shortages in water. Smallholder farmers use cattle for transport, animal-drawn ploughing, manure and as a symbol of wealth in most of the rural communities found in Southern Africa. However, in as much as the livestock provide these essential services for the rural community, water scarcity has been threating the livestock production in most areas where the ZRBF has been running the projects such as Mberengwa, Mbire and Sizemele (ZRBF, 2021a, b, c). Climate change shocks and threats have been affecting most rural communities in Southern Africa (United Nations Environment Programme, 2020). In line with this, ZRBF OMS 2 report (2020) notes that concurrent economic and natural shocks/threats are continuing throughout the country that aggravate the food insecurity situation and severely impact livelihoods, especially for poor households. Exposures to shocks and stresses have a strong causal relationship with a household's resilience capacities mainly due to the water scarcity that has been building up in most rural areas in Southern African countries (ZRBF OMS 2, 2020). Thus, because of this, it can be noted that water is an essential livelihood development product, and its scarcity has threatened the whole livelihood development value chain in most Southern African countries.

The above report also notes that the reduced flows and increased water stress have strong links to ecosystem health and water quality. This has been seen in most of the lowveld areas of Zimbabwe where fresh water has been a problem, leading to stunted growth in livestock and some crops in the area. In areas such as Mbire, Mberengwa, Mwenezi and Sikombela, water scarcity has been an issue that has been affecting the livelihood options of people living in the area. Hence, as noted by ZRBF OMS 2 report, there is greater need to mitigate the scarcity of water in these areas to boost resilience to climate change and develop more the livelihood options that are being carried out in the area. In line with this, the ZRBF OMS 3 report (2021c) notes that thus mitigating water scarcity contributes not only to greater physical resource availability but also to sustained ecosystem health and nearly all socioeconomic activities, including energy production, agriculture and food production and industrial activity.

4.3.2 Water-Based Adaptation and Resilience in Zimbabwe

Water adaptation challenge is a big problem in most Zimbabwean lowveld communities (ZRBF, 2020). In light of this, the overarching water management problem that requires a solution in the context of water management for adapting to climate change is the issue of having too little water in an area. This has been the problem that has been affecting the resilience programmes of people that are currently being run by the ZRBF in the lowveld areas of Zimbabwe. According to the ZRBF OMS 2 report (2020), because of the low rainfall that was received during the year, the resilience programmes of livestock and crop farming in the lowveld areas were heavily affected. As a result of this, the resilience programmes are facing various challenges due to unavailability of water, which is key in the running of these projects. However, shortages of water in many rural areas because of climate change has led to the development of water adaptation technologies. Considering this, UNFCCC (2014) notes that the specific technological means to execute the response/action on the ground include 'hard' technologies (e.g. construction of specific infrastructure, application of software, etc.) and 'soft' technologies (e.g. applying specific planning and management approaches and methods). These have been adopted by various projects being conducted by ZRBF in Zimbabwe, which have been run in the lowveld.

Buttressing this argument, the ZRBF OMS 2 report (2020) notes that as result of the threat to rural economies and livelihoods, the ZRBF embarked on programmes to increase water augmentation in the areas around Mberengwa and Nyanga in Zimbabwe. In line with this, the ZRBF OMS 3 report (2021c) notes that the water augmentation as an adaptation response aims to restore opportunities for increased water capture and infiltration by creating green spaces for improved natural recharge from rainfall. Another option for augmentation can be managed recharge, such as water injection in exiting aquifers. These options have managed to help in creating resilience of rural communities to water scarcity in their areas thereby managing their livelihood options and the various facets of rural economies in the area which has enabled them to stand against the water problems that climate change has posed against them in their communities. Supporting these arguments IPCC (2014) notes that the water adaptation that have been embarked on in most Southern African rural communities have gone a long way in cushioning these against the water problems that have been faced by these communities because of climate change. Table 4.1 below shows some of the water-based technologies.

Adapted from the UN Environment (2017:25) Report

Alternative strategies for addressing water scarcity include enhancing water efficiency and implementing demand management practices that mitigate the inefficient utilisation and wastage of freshwater through advanced technologies and improved regulation of water consumption (Climate Change Management Department, 2020). ZRBF OMS 1 (2019) indicates that anticipated alterations in the duration and intensity of dry and rainy seasons render water storage a critical adaptation strategy for addressing water scarcity. Certain regions may witness a reduction in precipitation, but others may experience an overall increase characterised by more intense rainfall over shorter durations, hence prolonging dry periods (McCartney & Smakhtin, 2010). Consequently, enhanced water allocation agreements are essential for the optimal utilisation of scarce resources and for guaranteeing improved coordination of resource development, thereby mitigating the danger of over-abstraction and ecosystem damage (IPCC, 2014). The ZRBF OMS 2 report elucidates that water allocation as a reaction to climate change seeks to adopt technology and build frameworks for the equitable distribution of current supplies among users, ensuring optimal benefits for all stakeholders, including the environment. The UNFCCC (2014) emphasised that in the context of resource scarcity, it is crucial to consider both the available supply and the demand that frequently contributes to water scarcity. Water efficiency and demand management strategies mitigate the inefficient utilisation and wastage of freshwater through enhanced technology and improved monitoring of water consumption (ZRBF OMS 3, 2021c). The reduction of water demand and usage in various areas must be concise and targeted to achieve effective outcomes. Despite the effective execution of water supply and demand strategies, certain places facing water scarcity necessitate the exploration of alternative water supply sources.

Table 4.1 Water-based technologies

Process	Technology	Description
Water allocation	Basin level modelling and seasonal forecasting	A planning instrument (or set of instruments) to help optimise water allocation in a river basin among competing water uses. A combination of hydrological and economic models is typically utilised, first to estimate total water availability, which can then be used to assess environmental and economic impacts of different allocation management scenarios in the basin. This can help decision-makers gain an overall picture of the current water use trends, and to make optimal (and sustainable) water management plans for the basin. A wide array of models is available
	Seasonal water rationing	An approach to control water use rates amongst different users based on the seasonal availability of water and socioeconomic priorities. Rationing may take the form of water use restrictions for certain purposes, in certain times or certain areas. The aim is to maintain equitable use among different users, as well as high levels of water productivity, throughout the year
	Water reallocation	The transfer of use rights between users who have been allocated a certain amount of water (through formal water use rights or entitlements, or informal arrangements), after it has been determined the initial allocation is physically impossible or socioeconomically unfavourable. Resource re-allocation can help adapt to unforeseen circumstances (e.g. critical water shortages during the dry season), reduce stress on renewable water supplies and help optimise water use benefits. Can either be voluntary or non-voluntary
	Rainwater harvesting for infiltration	Also known as in situ water harvesting, this is a practice in which rainwater uptake in soils is increased through the soil surface, rooting system and groundwater. Soil effectively acts as the storage agent, which improves water holding capacity and fertility and reduces risks of soil loss and erosion Examples include terracing, pitting and conservation tillage

4.3.3 Conceptual Framework: Water-Based Adaptation and Resilience

Water-based adaptation to climate change denotes the capacity of vulnerable individuals, households, communities and systems to foresee, mitigate, adjust, recover more effectively and progress from the impacts of shocks and hazards while safeguarding livelihoods and recovery achievements and fostering sustainable transformation reliant on water infrastructure at both household and community levels (ZRBF, 2019). The concept examines the resilience capacities of households and communities, which depend on the availability of water accessible to individuals, households, communities or systems, and how this resource is utilised to mitigate or avert the effects of shocks and stresses on developmental progress. Water-adaptive resilience is described as the capacity to coexist with shocks and stresses by leveraging the advantages provided by water supply. It entails modifications, selecting alternatives, adapting and evolving (for instance, implementing water conservation methods). The chapter ultimately delineates resilience pathways through water-based technologies, specifically examining the outcomes of employing resilience capacities and strategies that are crucial for sustainable livelihoods.

4.4 Methodology

This study employed qualitative data from the ZRBF, namely focus group discussions (FGDs) and key informant interviews (KIIs) from the OMS3 report, as it provides a comprehensive overview of the evolving tools over the years. However, prior surveys will be referenced to identify any modifications in water management practices. The ZRBF executed a qualitative study utilising a purposive sampling method. Qualitative sites were intentionally chosen from the clusters picked in the OMS3 reports for the quantitative analysis to examine significant characteristics of interest both across and within the clusters, spanning the seven consortia. The study established a set of selection criteria to encompass various ZRBF programming based on intervention type, level of uptake and proximity to an urban site, alongside programme intensity (high/medium) (ZRBF, 2020). In OMS3, with programme intensity (high/medium), the study established a selection criterion to encompass various ZRBF programming for intervention kind, uptake level and participation in off-farm or non-farm companies. In OMS 3, the qualitative research teams organised distinct focus groups for males and females, incorporating younger programme participants (under 35) to obtain insights and facilitate data analysis by gender and age. Discussions in focus groups (20 conducted) were generally varied for non-farm or off-farm enterprise groups, since the total number of members typically ranged from six to ten. Fourteen key informant interviews were done with individuals including lead farmers, community facilitators, Agritex officers and ward councillors. This chapter qualitatively examines numerous difficulties with OMS3. This encompasses shocks and their effects, focussing on age, gender and family type, the tactics employed to mitigate these shocks and pressures, as well as the utilisation and sources of information. The initiatives of ZRBF that assist individuals in managing shocks, particularly regarding water infrastructure, are very crucial in understanding water innovations (see ZRBF OMS3 Report page 140).

We employed a thematic analysis model methodology for data analysis. The thematic data analysis was multifaceted and predominantly inductive, allowing themes, topics, trends, patterns and conclusions to arise from the data entered into matrices. The thematic analysis of qualitative data, characterised by its intuitive and inductive nature, involved three distinct activities: (i) examining the data for themes, concepts and propositions; (ii) manually coding the data to enhance comprehension of participants' perceptions and (iii) identifying recurring themes across all matrices. The topics extracted from the qualitative data encompassed the efficacy of the water infrastructure, farmers' incentives to sustain the water infrastructure, obstacles to resilience and the characteristics of the resilience routes within the chosen consortiums. An examination of water activities at the consortium level and their influence on the ability to prepare for and adapt to environmental shocks and stresses was conducted (see pages 140–143 of OMS3). The research additionally utilises quantitative data from secondary sources, primarily OMS reports that were done across the various consortiums under ZRBF funding.

4.4.1 Selection of Consortiums

The chapter is based on both qualitative data and quantitative secondary data. The analysis was based on data drawn from three consortiums that were purposively selected. These are ECRAS, MELANA and ZVA. The different water innovations and technologies adopted in these particular areas are documented so as to understand the impacts of the ZRBF initiatives in these communities and how they have changed lives and livelihoods. Data abstraction from the three OMS reports indicates the following technologies were promoted by these consortiums (Table 4.2).

Table 4.2 Water-based practices and infrastructure promoted under ZRBF

OMS 1	OMS 2	OMS 3
Soil and water conservation techniques most familiar are minimum tillage (62% of the households use this), contour ridges/ contour planning (54%) Conservational agricultural practices (climate-smart agriculture) Compost organic Mulching Intercropping, plant diversity Drip/micro irrigation Integrated pest management\ minimum tillage Management and protection of watershed, planning of fodder trees	Water borehole solarisation provides more time for households to do other activities than fetching water manually Water point committee that repairs and maintains boreholes Water infrastructures such as dams, boreholes, weirs, garden drip irrigation schemes and household purchased water pumps Construction of water Harvesting and storage tanks improve livestock breeds; access to safe potable water is critical for livestock and agriculture and humans	Water point management such as fences and employing guards Monthly payment from 5 cents to 2 USD for the management of the infrastructure at water points Digging of trenches in waterlogged fields Use of river water to irrigate fields during drought seasons Training of communities in water conservation techniques such as drip irrigation, infiltration pit, dead level contours and mulching; drip irrigation had the highest use Water rationing also noted in some areas Borehole use (e.g. 128 boreholes sunk in Beitbridge through the Zimbabwe Resilience Building Fund; some are solar-powered) Community gardens Water harvesting

4.5 Results and Discussion

4.5.1 Water-Related Shocks and Stresses

Table 4.3 shows that in the ZVA consortium the susceptibility to various shocks and stress of climate change has been deepening over the years. The results show the mean susceptibility to shocks and stresses of the consortium from the OMS 1, 2 and 3 evaluation reports. From these statistics it can be noted that the susceptibility has been deepening over the years. Susceptibility to excessive rains has since grown from mean of 3.2 in round 1 to 9.2 in round 2 and 62.2 in round 3. The floods have also been increasingly affecting the area as seen through mean of 0.9 in round 1 and 4.8 in round 3 and 12.5 in round 3. Crop diseases or pest in ZVA consortium have been on the increase as noted in Table 4.3 from the mean of 27.3 in round 1 and 70.4 in round 2, whilst there is a decrease to 55.2 in round 3. Nevertheless, susceptibility to drought fell as noted in the results of OMS round 3 report with a mean of 66.2, 65.1 in round 2 and 31.5 in round 3. In addition to this, negative impacts to livelihoods of variable rains have significantly reduced as noted by the means of 50.1 in round 1, 72.1 in round 2 and 9.5 in round 3. These results show how water-related infrastructures in the area have managed to bring change in the area.

As presented in Table 4.4, it can be noted that in the MELANA consortium the susceptibility to various shocks and stresses of climate change has been deepening over the years. The results show the mean susceptibility to shocks and stresses of the consortium from the OMS 1, 2 and 3 evaluation reports. From these statistics it can be noted that the susceptibility has been deepening over the years. Susceptibility to excessive rains has since grown from a mean of 0.6 in round 1 to 3.7 in round 2 and 59.3 in round 3. The floods have also been increasingly affecting the area as seen through the mean of 0.0 in round 1, 0.6 in round 2 and 2.9 in round 3. Crop diseases or pest in MELANA consortium have been on the increase as noted in Table 4.4 from a mean of 26.7 in round 1 and 31.0 in round 2, whilst there is a decrease to 25.1 in round 3. Nevertheless, susceptibility to drought fell as noted in results of OMS round 3 report with a mean of 73.7, 83.5 in round 2 and 58.5 in round 3. Adding on to this, it can also be noted that the proneness to variable rains has also dropped from a mean of 78.9 in round 1, 58.3 in round 2 and 12.4 in round 3 of the

Table 4.3 Shocks and Stresses

ZVA consortium						
Variable	r1	r2	r1–2	r3	r2–3	r1–3
Excessive rains	3.2	9.2	+	62.2	***	***
Variable rains	50.1	72.1	+	9.5	***	***
Droughts	66.2	65.1	ns	31.5	***	***
Crop diseases or pests	27.3	70.4	***	55.2	+	**
Floods	0.9	4.8	+	12.5	ns	+

na not applicable; *ns* not statistically significant + $p < 0.1$ *$p < 0.05$ **$p < 0.01$ ***$p < 0.001$. Adapted from ZRBF OMS 32021 report, page 156

Table 4.4 Shocks and stresses experienced in MELANA consortium

MELANA consortium						
Variable	r1	r2	r1–2	r3	r2–3	r1–3
Excessive rains	0.6	3.7	**	59.3	***	***
Floods	0.0	0.6	ns	2.9	ns	+
Variable rains	78.9	58.3	***	12.4	***	***
Crop diseases or pests	26.7	31.0	ns	25.1	ns	ns
Droughts	73.7	87.5	**	58.	***	**

Adapted from ZRBF OMS 32021 report page 154

Table 4.5 Shocks and stresses experienced in ECRAS consortium

ECRAS consortium						
Variable	r1	r2	r1–2	r3	r2–3	r1–3
Excessive rains	5.9	4.7	ns	88.6	***	***
Floods	3.3	1.3	ns	9.9	+	+
Variable rains	85.6	75.2	**	28.9	***	***
Crop diseases or pests	66.1	64.3	ns	69.4	ns	ns
Droughts	25.1	70.9	***	58.2	+	***

Adapted from ZRBF OMS 32021 report page 153

OMS 3 studies. These results show how water-related infrastructures in the area have managed to bring change in the area.

As has been presented in Table 4.5, it can be noted that in the ECRAS consortium, the susceptibility to various shocks and stresses of climate change has been deepening over the years. The results show the mean susceptibility to shocks and stresses of the consortium from the OMS 1, 2 and 3 evaluation reports. From these statistics it can be noted that the susceptibility has been deepening over the years. Susceptibility to excessive rains has since grown from a mean of 5.9 in round 1 to 4.7 in round 2 and 88.6 in round 3. The floods have also been increasingly affecting the area as seen through mean of 3.3 in round 1 and 1.3 in round 2 and 9.9 in round 3. Crop diseases or pest in ZVA consortium have been on the increase as noted in Table 4.5 from a mean of 66.1 in round 1 and 64.3 in round 2, whilst there is a decrease to 69.4 in round 3. Nevertheless, susceptibility to drought fell as noted in the results of OMS 1 with a mean of 25.1, 70.9 in round 2 and 58.2 in round 3. Also, susceptibility to variable rains has fallen as shown by the mean of 85.6 in round 1, 75.2 in round 2 and 28.9 in round 3 of the OMS 3 studies. These results show how water-related infrastructures in the area have managed to bring change in the area.

In line with the above, one of the water-related shocks that was faced in consortiums was excessive rains in some of the seasons. In particular, in the ZVA consortium, excessive rainfall washed away fields and roads, some households lost their shelter as their huts were washed away as well. One participant explained that:

> *This season most people had actually planted quite bigger portions having seen better rains but as the season progressed their fields were then washed away. The crops were also affected but the lack of fertilizers as they fields were waterlogged and had the fertilizers that had been applied leached.* ZRBF, OMS 3 Report of 2021c

As a result, the ZVA consortium was plunged into deep water problems which were at the core of the poverty, affecting most people in the community. In the FGDs conducted some of the participants highlighted that "*we are now short supply of fresh water for drinking which is now affecting our health*". In addition to this, as was noted in the FGD matrix reviewed, the excessive rains also led to the breeding of many mosquitoes in the areas, and they have seen a surge in cases of malaria in their community. Some people have been affected by worms that are found in the water as most people walk barefoot. Because of the heavy rains that were experienced that season, some community members lost some goats to a disease that they experienced, where the goats lost so much weight, and their condition deteriorated quickly, leading to their death.

Another challenge that has been faced in the consortium of MELANA is the issue of drought. Drought has made the water table to go down in the area, making the areas in Sub-Saharan African countries prone to more water shortages (IPCC, 2014). As was noted in the FGD matrices, most participants highlighted that the "*the previous season 2019/20 we experienced a drought on the back of other droughts in the preceding years such that the crops we planted in that season failed leaving most people without food stocks*" (ZRBF OMS 3 Report FGD extracts). As a result of the drought which brought in more water shortages in the area the pasture, the establishment was severely affected to the extent that most people had to sell off their livestock at very low prices just to get a bit of return than completely lose out, as the livestock died. Traditionally people in the consortium would also rely on relief aid which was brought by different NGOs, but in the period under review, this was not possible due to travel restrictions that had been instituted in response to the COVID-19 pandemic. Also, in the consortium of MELANA, as a result of water shortages, the market gardening that was done in the area has also gone down. In the FGDs, the people in the area usually undertook gardening, but in the previous season, they were not able to establish these gardens as there was no adequate water to irrigate the plants in the gardens. This was mainly so because the wells they use to water the gardens dried up early that season, which left them without the income they normally get from selling the vegetables as well as the diverse food reserves they depended on. As a result, the shortages of water largely affected the people in MELANA as their capacities and production have been reduced due to the heavy water shortages experienced in the previous seasons.

The lack of water due to the drought they experienced in the 2019/20 season, affected water availability for both domestic use and livestock in the ECRAS consortium. In as much as ECRAS consortiums had boreholes in the area, these dried up early and could not meet the demand for water by the growing population in the community. As noted in the FGDs conducted, some participants highlighted that:

> *People would go out as early as 2am to queue for water and try to draw water before most people come, having gone out that early at times one would then only be able to get water at 10am and be limited to 3 buckets and this water was meant for both domestic use and for their livestock.* ZRBF OMS 3 2021 Report FGD Source

Consequently, the water rationing system managed to make more people access water, but the water was not enough for their livestock, crops and other domestic processes. In addition to this, another challenge faced was on the quality of the water they obtained from the boreholes. The water was salty and this at times caused running tummies in children, so they ended up having to go to springs which are far to fetch fresh water for domestic use. Furthermore, drought also resulted in livestock deaths since there was no forage in the grazing lands. The boreholes that are found in the ward dried up such that people had to use water from the Save River for domestic purposes and livestock. This caused diarrhoea and bilharzia in people. In line with this, McCartney and Smakhitin (2010) notes that the shortage of the water in most water-prone areas in Sub-Saharan Africa has led to the rise of diseases that are water borne.

4.5.2 Water-Related Infrastructure: Resilience Capacities

The three consortiums in this study have some form of water-based infrastructure that have been developed. From the OMS 3 data, it was noted that in ECRAS, there are more water-based infrastructures as compared with others (it has forms of diversified water infrastructure, which include boreholes, water harvesting facilities, irrigation and mechanised water facilities). The ECRAS Consortium is then followed by ZVA, which has three forms of diversified water infrastructures, such as boreholes, water harvesting facilities and a dam, though it is silted. The MELANA consortium, though it had water-based infrastructure, has the least of these with only two forms of water infrastructure, which are the boreholes and irrigation. Water harvesting is one of the mechanisms that is being used to respond to water challenges faced in ZVA, ECRAS and MELANA. Water harvesting in Africa has proved handy in various communities as it has helped in agricultural use and domestic use in many households (IPCC, 2014). The water is harvested from the rooftops into the set tanks and is used for gardening and other home processes. Gardening and livestock drinking are the main uses for the harvested water. In ZVA and ECRAS, the harvested water was useful since it helped in some of the livelihood options for the household and community, such as gardening, livestock drinking and basic domestic use, which forms the basis of the income generation for the household.

Drilling of boreholes was one of the ways that consortiums were using to address the water challenges that they are facing. In as much as the lowveld region generally experiences water scarcity, this challenge has continued to increase with climate change. According to the AGRITEX officer "*Through ZRBF and their projects, it was noted that drilling of boreholes is being implemented in order to try and help the communities to be water secure*". ZVA, ECRAS and MELANA drilled boreholes. T ECRAS was reported to have a viable management committee of the boreholes found at community level. The drilling of the boreholes assisted in helping the communities in their agriculture endeavours such as livestock production and crop cultivation. Drilling of boreholes also helped in the provision of water for livestock such as the cattle, goats and chicken. The water is also used in community dip tanks, which also help in reducing livestock diseases.

Some of the boreholes were mechanised, especially in the ECRAS consortium. The mechanised water facilities include solar-powered boreholes, irrigation channels and tapped water in households (ZRBF, OMS 32021). In line with this, one participant explained that "*The mechanized water facilities have reduced the distance we used to walk to fetch water and the challenges of having back aches and other load carrying body problems*". Mechanisation enabled the communities to develop various water-based livelihood options such as market gardening and livestock production at household level, thereby enabling them to become more and more food secure. The mechanised water facilities have reduced the labour needed to fetch water by families and communities, thereby making them water secure (ibid). The maintenance of the facilities is community based, and this has made the community to take ownership of the mechanised boreholes and water points in the communities. Through the payment of US$1 as subscription and maintenance fee, this has given the community power to summon repairers and do mandatory checks and repairs to their water sources, which has given them a source of ownership. Selected consortiums have ventured into aquaculture in fishponds. The fish have greatly helped in the diversification of diets of the people in the ZVA and MELANA consortiums. In the FGDs conducted most participants explained that "*through such endeavours the communities have also found other livelihood options which has helped in income generation for them and their families*". The fishponds have acted as replicas of dams but at a small scale, which makes them easy to manage when dealing with fish farming in the lowveld (ZRBF, 2020).

4.5.3 Motivations for Supporting Water Infrastructure

From the review of the data from OMS 3 reports, it was noted that one of the key motivations for supporting water infrastructure by ZRBF was the need for access to water for people and animals in the communities. Since most consortiums occur in the lowveld, which is prone to high water accessibility insecurities, the establishment of water infrastructure in these communities improved access to water. In the FGDs conducted, it was noted that in the communities of ZVA and ECRAS, the installation of water infrastructure helped greatly in creating better accessibility to water. UN Water (2019) notes that access to water means access to food and stable income in these communities. Another motivation for the establishment of the water infrastructure in the communities was the need for income generation. The lowveld is mostly dry with low rainfall, making it difficult to conduct any meaningful diversified livelihood without water. Therefore, the establishment of various forms of water infrastructure was done with the motivation and need to diversify the livelihood of people amidst the climate change impacts that have been affecting the community greatly. The establishment of the water-based infrastructure would also help to generate income and help the communities to move out of the clutches of poverty

(UN WATER, 2019). Through the water-based infrastructure, communities have now been able to develop market gardening initiatives, fish farming and small livestock production, as was noted in the OMS 3 report. The other motivation for developing water infrastructure was the need to develop a healthy community through provision of clean and fresh water. OMS 3 report notes that some of the consortiums were facing problems of water-shortage related diseases such as cholera, typhoid and bilharzia in their communities. As such, the drilling of boreholes helped greatly in reducing the incidences of water borne diseases that were being experienced in the communities (ZRBF OMS 3 Report, 2021c).

4.5.4 Resilience Outcomes from Water-Based Infrastructure

Table 4.6 shows that through the introduction of water infrastructures in MELANA, the communities and households have now been able to be resilient to the identified shocks and stresses. In the comparisons of ZRBF OMS 1, 2 and 3 reports, it can be noted that resilience outcomes have been largely positive. MELANA consortium has become resilient to excessive rains, as noted, through having the mean of 27.0 in round 3, which is a significant improvement form OMS 1 and 2. In terms of the drought, it was noted that the resilience had improved as shown by the mean of 4.2 in round 1, 28.1 in round 2 and 50.5 in round 3. When it comes to crop and livestock diseases, resilience had improved, as noted, through 5.7 in round 1, 28.5 in round 2 and 17.6 in round 3. When it comes to resilience to variable rains, it was noted that the resilience had improved through the mean of 3.4 in round 1, 17.5 in round 2 and 63.3 in round 3. The statistics clearly show how the interventions have brought about resilience outcomes in the communities.

In addition to this, Table 4.7 shows the changes in the outcomes that have happened in ECRAS in response to climate change impacts. The table shows that the climate changes has affected the environmental set-up of the area, which has a

Table 4.6 Changes in outcomes in MELANA consortium

MELANA consortium									
Variable	r1	r2	r1–2	r3	r2–3	r1–3	r1	r2	r3
Excessive rains	na	na	***	27.0	3	6	170		
Variable rains	3.4	17.5	***	63.0	***	***	329	222	44
Drought	4.2	28.1	***	50.5	***	***	289	361	223
Crop diseases or pests	5.7	28.5	**	17.6	ns	+	94	103	67
Livestock diseases	na	na	ns	na	15	14	8		
Sharp food price increase	2.1	12.5	**	25.2	**	***	380	269	232
Death of household member	na	na	na	19	16	11			
COVID restrictions	41.0	225							

Adapted from ZRBF OMS 32021 report 159
na not applicable; *ns* not statistically significant + $p < 0.1$ *$p < 0.05$ **$p < 0.01$ ***$p < 0.001$

Table 4.7 Changes in outcomes in ECRAS consortium

ECRAS consortium									
Variable	r1	r2	r1–2	r3	r2–3	r1–3	r1	r2	r3
Excessive rains	na	na	65.4	16	10	356			
Variable rains	69.5	46.2	***	80.5	***	*	375	318	133
Drought	48.4	47.6	ns	73.6	**	**	111	314	269
Crop diseases or pests	66.2	49.0	**	70.3	***	ns	303	254	302
Livestock theft	40.5	57.6	ns	45.6	ns	ns	40	39	38
Sharp food price increase	62.7	37.1	***	63.1	***	ns	380	375	228
Death of household member 59.6	na	na	25	18	16				
COVID restrictions	68.7	324							

Adapted from ZRBF OMS 32021 report 158

Table 4.8 Changes in outcomes in ZVA consortium

ZVA consortium									
Variable	r1	r2	r1–2	r3	r2–3	r1–3	r1	r2	r3
Excessive rains	na	17.4	13.6	ns	ns	15	30	261	
Variable rains	10.3	11.7	ns	22.2	ns	ns	224	300	36
Drought	9.8	6.3	ns	17.7	+	ns	302	284	146
Crop diseases	20.7	14.7	ns	21.5	ns	ns	112	263	230
Livestock theft	na	17.2	18.3	ns	12	37	43		
Sharp food price increase	10.4	9.7	ns	17.0	ns	+	297	316	226
Death of household member	na	na	na	21	17	17			

Adapted from ZRBF OMS 32021 report 159

bearing on water and water-based lives and livelihoods in the community. However, in as much as this has been noted, the table emphasises on the resilience levels of the consortium, which have positively changed over time. Amidst the problems being faced, the ability to bounce forward, adjust and adapt has been seen to be growing as highlighted in the table.

From Table 4.7, comparing ZRBF OMS 1, 2 and 3 reports results, it can be noted that resilience outcomes have been largely positive. ECRAS consortium has become resilient to excessive rains (highly significant mean of 65.4.0 in round 3 compared to OMS 1 and 2). Resilience to drought improved (mean of 48.4 in round 1, 47.6 in round 2 and 80.5 in round 3). Resilience to crop diseases and pests also improved (66.2 in round 1, 49.0 in round 2 and 70.3 in round 3). Resilience to variable rains had a mean of 69.5 in round 1, 46.2 in round 2 and 80.5 in round 3, showing an improvement. The statistics highlight how the interventions improved the resilience of communities to various shocks and stress in ECRAS.

Table 4.8 shows resilience outcomes have been largely positive from OMS 1 to 3. ZVA consortium became resilient to excessive rains (mean of 17.4 in round 3 compared to a mean of 13.4 in OMS 2). Communities became resilient to drought (mean of 9.8 in round 1, 6.3 in round 2 and 17.7 in round 3). Crop and livestock diseases resilience improved (20.7 in round 1, 14.7 in round 2 and 21.5 in round 3) as well as resilience to variable rains (mean of 10.3 in round 1, 11.7 in round 2 and 22.2 in round 3). The statistics clearly show how the interventions caused statistically significant improvement in resilience in the communities.

Qualitative data showed that the mechanised water facilities introduced in ECRAS, ZVA and MELANA improved water management in these consortiums. The introduction of solar-powered boreholes, tapped water and irrigation facilities ensured that women and children put in less labour in fetching water for use at the household level. The mechanised water facilities greatly helped in making sure that the livelihood options for the people are well developed as the water facilities have given them the power to venture into market gardening, fish farming and many other income-generating activities (ZRBF OMS 3 report, 2021c). The mechanised water facilities incentivised the livelihood options that were already existing in the communities and have widened the income-generating activities of the people in the communities. The people managed to diversity their diet and income-generating activities. Some of the outputs of the water infrastructure installed in the communities include good vegetable and fruit production, healthy livestock such as goats, chickens and cattle and also fish, which has made them food secure.

The introduction of water facilities helped in improving the health of the people. As was noted in the interview conducted with an AGRITEX Officer in ZVA, he explained that "*before the introduction of water facilities communities under ZVA and ECRAS consortiums were susceptible to diseases such as cholera, bilharzia and typhoid due to lack of access to clean water*". The incidences of waterborne diseases declined once water became available (see ZRBF OMS 2 Report, 2020). Water harvesting and introduction of water points helped the women and children in the consortiums greatly. In the FGDs conducted with women in ECRAS, some participants explained that "*the water points and water harvesting technologies have relieved the women found in the consortiums of the burden of fetching water from long distances as it was challenge for most of them*". Through these water provisioning methods, the burden of searching for water and carrying the water for the households has been removed, which has been so helpful in making sure that the women have time to do some other income-generating projects for their families (ZRBF, 2020 OMS 2 Report, 2020). The availability of water near households has enabled women to venture into other income-generating ventures such as vending, sewing and other livelihood options (OMS 3 report 2021c).

The construction of fishponds and irrigation schemes in ZVA, MELANA and ECRAS consortiums enabled the communities to diversify their livelihood options. In the interviews conducted with one village head in ECRAS, he explained that "*setting up of irrigation schemes such as Gudo Irrigation Scheme has enabled the communities to conduct market gardening. The communities have managed to produce vegetables and fruits such tomatoes, covo and other horticultural produce such as butternuts as result of the establishment of the irrigation schemes*". The irrigation facility has revitalised the businesses of many people in the consortiums, as these now can have extra income from farming whilst at the same time getting diversified diet at the family level. The introduction of the fishponds for fish farming has also been beneficial to the communities. Selling fish provided additional income at the household level and introduction of fish in diets has diversified the diet at the household level. Most households within the three consortiums managed to diversify their diet and to earn extra income because of the fish farming and irrigation schemes.

The setting up of water infrastructure also helped the communities to become resilient to climate change (IPCC, 2014). The drilling of boreholes, installation of mechanised water facilities and rain harvesting technologies helped communities under the three consortiums to adapt to the climatic changes that had caused water shortages due to the lowering of the water tables and drying up of already silted rivers. The livelihood options for the households under the ECRAS, ZVA and MELANA consortiums had been negatively affected. Participants in one FGD conducted in MELANA echoed this by explaining that "*in terms of livestock, cattle and goats were now becoming stunted in their growth whilst crop cultivation and yields dropped to very low levels as a result of water shortages*". Thus, the introduction of water facilities in the community helped greatly in restoring agriculture. Crop cultivation and the livestock production was boosted, and profits were realised. In market gardening, selected households were realising between US$50–100 per day after selling their produce. Water facilities installed in the consortiums enabled people to generate income for the family's basic needs.

The establishment of water infrastructure enabled the communities to have meaningful income-generating plans (ZRBF OMS 3 Report, 2021c) through fish farming, crop cultivation and livestock. In the ZVA consortium, one respondent explained that "*through the establishment of water facilities we have managed to have embark on fish farming and market gardening which has given us extra income. Now monthly we can sustain our families as we are getting income*". Households thrived on selling their produce such as fish, tomatoes, goat meat and milk and chickens. The income generated was used for the payment of school fees, securing food and payment of medical bills by the beneficiaries in the communities.

4.6 Challenges to Resilience

Activities such as gold panning and brick moulding, which also used water from some of the facilities set up as water stations in the communities, were considered as causing environmental degradation. In the FGDs conducted, some participants noted that "*the processes of gold panning and brick making that has been occurring in MELANA consortium has left the area environmentally degraded as gullies have been formed in the area which have become a hazard*". Brick making and gold panning in Bubi and Binga left behind trials of gullies in the communities which have become a harm to both people and animals (ZRBF OMS 3 Report, 2021c). This has greatly affected the environmental conditions of the areas in which the activities are being carried out, thereby affecting sustainability of the environment.

Another challenge that may affect the sustainability of the efforts made is that water infrastructure projects and water facilities development was on a very small scale such that many still have no access. Not all wards have boreholes and water points. One village head explained that "*One of the challenges that we have is that though we have boreholes in MELANA consortium some of the villages and wards still do not have these boreholes whilst some of the boreholes drilled have water that is very much high in lime which makes it dangerous to consume*". The provision of water infrastructure that is not widely inclusive is problematic as those that are left behind may still continue to be food insecure and also prone to water challenges (UNFCCC, 2014). There is a need for holistic approaches to provision of water facilities so that the impact of the water provision is significant.

The introduction of water facilities also increased competition for produce in the communities. This affected the pricing of the produce such as tomatoes, goat meat, chickens and milk. Due to competition in the pricing of produce from activities such as crop cultivation and livestock rearing, sustainability of these now has become highly questionable (ZRBF OMS 2 Report, 2020). The competition for markets became stiff such that some farmers opted to sell their produce at non-profitable prices, especially perishables. This left some of the farmers broke since they were not realising much profits. The stiff competition in the market for produce led to the unsustainability of the farming activities for farmers. Communities paid maintenance fees for the upkeep of water infrastructure (e.g. boreholes). In MELANA consortium, it was noted that "*members of the consortiums pay an amount equivalent to US$1 per household each on a monthly basis that would be collected in case of the need for repairs of the boreholes and water facilities*". The repairs for the water facilities can be costly depending on what would have been damaged in the process, thus the amounts being charged at the consortium level may be insufficient, if added together, to do normal repairs and servicing of the boreholes. The cost of normal repairs and servicing of the water facilities, which are higher as compared to the low subscription fees that are being paid by the communities, may lead to the unsustainable running of water management projects (ZRBF OMS 2 Report, 2020). The high cost of repairs for the water facilities in comparison to subscription fees and levies evokes the question of sustainability for the water facilities' repairs and functioning in the community.

In as much as the crop cultivation and livestock livelihood options are growing in the ECRAS, ZVA and MELANA consortiums, these have been threatened by the livestock diseases and crop diseases. In MELANA it was noted that livestock diseases such as black leg and lumpy skin for cattle were prevalent (ZRBF OMS 3, 2021c). Ticks have also been a problem for goats and cattle. In the FGDS conducted some of the participants explained that "*crops such as sorghum, millet and rapoko are also being affected by qualia birds which have greatly reduced our crop yield thereby making our families much more prone to poverty*". Thus, the rise of diseases and pests for both crops and livestock has threatened the sustainability of the endeavours undertaken by the consortiums. In as much as water is now readily available for livestock and crops, this positive gain on livelihoods is however being threatened and is now being posed on the livelihood options. Though water harvesting is being used to help manoeuvre around issues of water challenges, this is a seasonal activity. This necessitates the need to use other means and ways to manoeuvre around water challenges.

4.7 Conclusion

Climatic shocks and stresses are real in rural communities covered by the consortiums discussed in this chapter. Droughts continue to threaten the availability of water for domestic use, livestock and crop production. The stresses and shocks are intertwined. As water is contaminated, disease outbreaks also occur, affecting human health. Droughts have also contributed to forage stresses and farmers are forced to sell their livestock, usually below the market value. Water shortages also affect horticulture and gardening which then affect dietary diversity, food security and incomes. The main conclusion is that resilience capacities in the form of water infrastructure have in some way contributed to resilience. Improved health from diversified diets has been noted. Improved food security and incomes also emerged in the consortiums. There is also improved livestock health due to availability of water. All in all, the various resilience capacities lead to different outcomes, but also converge to produce multiple outcomes. For instance, fishponds have led to fish farming from which healthy diets, improved food security and incomes have resulted. But these also result from irrigation, which allows crop diversification in gardens.

References

Caretta, M. A., Mukherji, A., Arfanuzzaman, M., Betts, R. A., Gelfan, A., Hirabayashi, Y., Lissner, T. K., Liu, J., Lopez Gunn, E., Morgan, R., Mwanga, S., & Supratid, S. (2022). Water. In H.-O. Pörtner, D. C. Roberts, M. Tignor, E. S. Poloczanska, K. Mintenbeck, A. Alegría, M. Craig, S. Langsdorf, S. Löschke, V. Möller, A. Okem, & B. Rama (Eds.), *Climate change 2022: impacts, adaptation and vulnerability. Contribution of working group ii to the sixth assessment report of the intergovernmental panel on climate change* (pp. 551–712). Cambridge University Press. https://doi.org/10.1017/9781009325844.006

Fecht, S. (2019). *How climate change impacts our water*. Available at: https://news.climate.columbia.edu/2019/09/23/climate-change-impacts-water/

Hajkowicz, S., & Higgins, A. (2008). A comparison of multiple criteria analysis techniques for water resource management. *European Journal of Operational Research, 184*, 255–265.

IPCC. (2014). Freshwater resources. In B. E. Jiménez Cisneros, T. Oki, N. W. Arnell, G. Benito, J. G. Cogley, P. Döll, T. Jiang, & S. S. Mwakalila (Eds.), *Climate change 2014: Impacts, adaptation, and vulnerability. Part A: global and sectoral aspects. contribution of working group II to the Fifth assessment report of the intergovernmental panel on climate change* (pp. 229–269). Cambridge, United Kingdom and New York, NY, USA, Cambridge University Press.

IPCC. (2021). Climate Change. In *The physical science basis*. Cambridge University Press.

McCartney, M., & Smakhtin, V. (2010). *Water storage in an era of climate change: Addressing the challenge of increasing rainfall variability*. International Water Management Institute.

Traerup, S. L. M., & Bakkegaard, R. K. (2015). *Evaluating and prioritizing technologies for adaptation to climate change – a hands on guidance to multi criteria analysis (MCA) and the identification and assessment of related criteria*. UNEP DTU Partnership. http://www.tech-action.org/Publications/TNA-Guidebooks

UN Water. (2019). *Climate change and water*. UN Water Policy Brief. Available at: https://www.unwater.org/app/uploads/2019/10/UN_Water_PolicyBrief_ClimateChange_Water.pdf

UNEP. (2020). *Climate change adaptation technologies for water*. UNEP-DHI Centre.

UNESCO, UN-Water. (2020). *United Nations world Water development report 2020: Water and climate change*. UNESCO.

UNFCCC. (2014). *Technologies for Adaptation in the Water sector*. TEC Brief #5, United Nations Framework Convention on Climate Change.

Urama, K. C., & Ozor, N. (2010). Impacts of climate change on water resources in Africa: The role of adaptation. African technology policy studies. *Network, 29*.

Xia, R., Zhang, Y., Critto, A., Wu, J., Fan, J., Zheng, Z., & Zhang, Y. (2016). The potential impacts of climate change factors on freshwater eutrophication: Implications for research and countermeasures of Water Management in China. *Sustainability, 8*(3), 1–7.

ZRBF. (2019). Outcome monitoring survey round 1. In *Programme learning report*. ZRBF.

ZRBF. (2020). Outcome monitoring survey round 2. In *Programme learning report*. ZRBF.

ZRBF. (2021a). *Zimbabwe resilience building fund humanitarian assessment report 2021 program learning report*.

ZRBF. (2021b). Outcome monitoring survey round 3. In *Consortiums qualitative fieldwork data 2021*. ZRBF.

ZRBF. (2021c). Outcome monitoring survey round 3. In *Programme learning report*. ZRBF.

Chapter 5
Building Asset Stock for Resilience in Vulnerable Communities in Zimbabwe

Emmanuel Mwakiwa, Edward Mutandwa, Pamela Madududu, Charity Mumaniki, and Simbarashe Tatsvarei

5.1 Introduction

Developing countries are afflicted by natural shocks such as droughts, floods and storms (Franklin, 2019). The effects of such disasters are exacerbated when shocks simultaneously affect members of a community. In addition to natural disasters, exogenous shocks to income and consumption disrupt poorly equipped low-income households. As households in the developing world are exposed to new or increasingly severe climate and economic shocks, pro-poor programmes have prioritised building household resilience (World Bank, 2016; Food and Agriculture Organization et al., 2018). Lack of adequate, suitable and affordable insurance arrangements puts households at greater risk of shocks (Dercon et al., 2005; Jensen & Barrett, 2017). Because of various climatic and economic shocks affecting livelihoods and the economy, the World Development Report (2014) expressed the need to manage risks as an important pathway for lessening vulnerability and reinforcing resilience for economic growth and development.

E. Mwakiwa (✉) · E. Mutandwa · S. Tatsvarei
Department of Agricultural Business Development and Economics, Faculty of Agriculture, Environment and Food Systems, University of Zimbabwe, Harare, Zimbabwe
e-mail: emakiwa@agric.uz.ac.zw

P. Madududu
Department of Agricultural Economics, Extension and Rural Development, Faculty of Natural and Agricultural Sciences, University of Pretoria, Pretoria, South Africa
e-mail: madududu.p@tuks.co.za

C. Mumaniki
Economics, Markets and Trade Department, Ministry of Lands, Agriculture, Fisheries, Water and Rural Development, Harare, Zimbabwe

L. Mhlanga et al. (eds.), *Building Resilience to Climate Change in Selected Districts in Zimbabwe*, https://doi.org/10.1007/978-3-031-80529-5_5

There is increasing policy interest on the impacts of shocks on welfare outcomes and asset building in developing countries (Dercon et al., 2005; Manlosa et al., 2019). Asset building relates to how individuals, families and communities gather the resources to move them towards economic well-being, for now and for years to come (World Bank, 2016). The assets include tangible capitals (natural, physical and financial) and intangible capitals (social and human) (Manlosa et al., 2019). These assets include those held individually or in a household (a house or consumer durables, labour, accumulated wealth, education and health) and those held by community, such as infrastructure, services, natural resources, accrued wealth in community savings groups and community organisations' social capital. The central role that assets can play in poverty reduction and development is widely recognised in the literature (Jabeen, 2015; Eichsteller et al., 2022). However, recognising the role of assets in supporting smallholder vulnerable communities and households to build their resilience to shocks remains implicit (Prowse & Scott, 2008; da Costa et al., 2013).

5.1.1 Vulnerability to Livelihood Shocks Among Zimbabwean Smallholder Farmers

Smallholder farmers in Zimbabwe have had their welfare adversely affected by various shocks. The shocks include droughts and floods, crop and animal diseases, economic shocks such as price inflation of food and farming inputs and loss of employment (Cornelius & Kiiza, 2021; Mugandani et al., 2012). In addition, extensive poverty and food insecurity, high levels of HIV/AIDS and persistent economic challenges have created vulnerable conditions (Cesvi, 2019).

Against this background, the Zimbabwe Resilience Building Fund (ZRBF) was promulgated to address several severe economic, environmental and social shocks and stresses affecting rural communities in Zimbabwe (UNDP, 2020). Building resilience is about enhancing and reinforcing the capacities, livelihoods and opportunities of the most vulnerable and food-insecure people, communities and countries in the face of an increasingly risky environment (WFP, 2011; Mwadzingeni et al., 2022). From a social-ecological perspective, resilience is conceptualised as the capacity of socio-economic systems (e.g. households) to withstand shocks through absorption, adaptation and transformation (UNDP, 2020). The ZRBF Theory of Change is anchored on the hypothetical proposition that if investments are made to

(i) Directly support targeted communities to improve absorptive, adaptive and transformative capacities
(ii) Avail timely and cost-effective responses to emergencies
(iii) Generate learning and utilise evidence in policies and decisions

Then, not only will targeted communities be more resilient and food secure, but there will also be a better understanding of what works and what does not work to strengthen communities' resilience in Zimbabwe (UNDP, 2020).

5.1.2 The Role of Assets in Building Resilience

An asset generates flows or consumption and additional stock (Ford Foundation, 2004). Assets can be a household's wealth stock, and this can be extended to include intangibles such as household relations, human capital and social entitlements (Sherraden, 1991; Manyumwa et al., 2018). In the contemporary poverty-related development discourse, the concept of assets or capital endowments is often recognised as critical in rural development (Moser, 2006). Despite the issues arising with the capital asset framework's roles in disaster recovery, few research studies have used capital assets to determine household vulnerability (Zhang et al., 2019). In addition, there is not much available data from disaster situations, which includes measuring of the five types of capital assets (Zhang et al., 2019).

In the context of low-income local communities and households, there is already much evidence on the actions taken to cope with shocks (Marzi et al., 2019). Coping mechanisms are distinguished by being reactive and geared towards survival, while adaptation involves planning with both shorter- and longer-term outlooks and, therefore, should contribute to, rather than deplete, assets (CARE, 2009). However, the extent to which adaptation can be done by vulnerable individual households or smallholder communities is limited (CARE, 2009). At the community level, systems and paths need to be included to the broader trunk infrastructure, and climate impacts should be determined locally by systems managed more broadly, such as watersheds (Hendriks, 2015). At the individual household level, assets lead to improvements in resilience capacity, which leads to better responses by households, which in turn helps them maintain or improve their food security or other well-being outcomes in the face of shocks or stresses (Hendriks, 2015).

5.2 Research Objectives

The major purpose of this research was to assess the relationship between assets and resilience among rural households in Zimbabwe.

The following were the specific objectives of the research study:

(i) To ascertain types of shocks faced by households
(ii) To determine the types of interventions in the project areas
(iii) To examine how smallholder vulnerable communities in Zimbabwe plan for and respond to external shocks

(iv) To examine the extent to which household resilience capacity is influenced by livelihood assets
(v) To establish the factors affecting household resilience

5.3 Research Questions

In particular, the study sought to answer the following research questions:

(i) What types of shocks prevail in rural settings?
(ii) What are the different types of interventions that are being implemented in the project area?
(iii) What coping and adaptive mechanisms do households with different socio-economic characteristics employ to manage economic, environmental and health shocks that disrupt their asset holding?
(iv) To what extent do livelihood assets influence household resilience capacity?
(v) What are the socio-economic factors that affect household resilience to shocks?

5.4 Methodology

5.4.1 Data and Data Sources

Using panel data from three waves of outcome monitoring surveys (OMS) conducted by the ZRBF, this chapter aims to understand how smallholder vulnerable communities in Zimbabwe plan for, and respond to, external shocks so as to identify approaches to promoting more sustainable, inclusive and resilient rural development. The research seeks to determine the extent to which adaptive or coping actions that use existing assets at the household or community level can be used in cooperation interventions between the communities and other stakeholders, including local government, local NGOs and other institutions, to reduce the current vulnerability to risk.

The study used 3 waves of panel data from the OMS conducted by the ZRBF from 2019 to 2021 from 18 districts in rural Zimbabwe from 7 consortiums, which are as follows:

- Building Resilience through improving the Absorptive and Adaptive Capacity for Transformation (BRACT) (Mutoko and Mudzi)
- Sizimele Action for Building Resilience in Zimbabwe (SIZIMELE) (Matobo, Insiza and Lupane)
- Enhancing Community Resilience and Sustainability (ECRAS) (Chiredzi and Mwenezi)

- Enhancing Community Resilience and Inclusive Market Systems (ECRIMS) (Mberengwa and Zvishavane)
- Matabeleland Enhanced Livelihoods Agriculture and Nutrition Adaptation (MELANA) (Bubi, Nkayi, Umguza and Umzingwane)
- Program for Growth and Resilience (PROGRESS) (Beitbridge and Nyanga)
- Zambezi Valley Alliance for Building Community (ZVA) (Binga, Kariba and Mbire)

ZRBF commenced in 2015 and was extended to September 2022, supporting the implementation of resilience-building activities in the 18 districts. Data on 11 ZRBF topline indicators were collected, including outcome variables (e.g. use of the practice of project-supported activities) and impact variables (e.g. household income, food security and nutrition) (UNDP, 2020). The survey design was stratified into the 'high-programming intensity' and 'medium programming intensity' categories based on the number of specific project activities each beneficiary had engaged in, according to project records (UNDP, 2020). Beneficiaries with two or less programme intervention exposures were categorised as 'Medium intensity', while those with three or more exposures were categorised as 'High intensity'. Households' livelihood strategies were characterised by household cash sources and food sources.

5.4.2 *Conceptual Framework for Assessing the Relationship Between Household Resilience Capacity and Livelihood Assets*

The ZRBF programme is one of the numerous initiatives that were enunciated by civil society to increase household and community resilience. The programme focuses on approximately 800,000 households, predominantly located in the country's arid and semi-arid regions. The programme was anchored on three interrelated components enumerated below:

- Component 1: Building evidence to improve the policy environment and stimulate service provision to enhance household and community resilience
- Component 2: Interventions that support long-term household and community resilience that focus in particular on building absorptive, adaptive and transformative capacities
- Component 3: A risk financing mechanism to make timely, appropriate and predictable fundings available for target communities that experience humanitarian shocks

The collective impact of the above-mentioned objectives is households' capacity to withstand natural stresses and shocks from unpredictable climate change events. In this chapter, we focus on the contribution of the ZRBF to building rural household resilience. In particular, the focus was on how livelihood assets are used to generate sustainable resilience capacity flows for the household. Resilience capacity is an expression of the household's absorptive, adaptive and transformative domains (Kebede et al., 2016). These domains are, in turn, affected by the resources the households have access to in the face of naturally induced shocks. Individual households cannot withstand the effects of exogenous events without assets. Resilience is a broad-based process that can be observed over temporal and spatial scales (Asmamaw et al., 2019). In this chapter, we adopted the ZRBF resilience indexing methodology. Under this approach, household resilience is an amalgamation of absorptive, adaptive and transformative capacities based on numerous socio-economic and demographic variables (ZRBF XE "Zimbabwe Resilience Building Fund (ZRBF)" , 2022). The resilience index is a percentage with higher values denoting a higher capacity to absorb, adapt and transform against climate change.

To assess the relationship between resilience capacity and underlying socio-economic and demographic predictors, we are guided by the Disaster Resilience Framework (DRF) (Fig. 5.1).

The disaster resilience framework is constituted by four elements defining resilience, which are the context, disturbance, the capacity to deal with disturbance and reaction to disturbance (FISN, 2014). The resilience model can be applied to individual, household, community and national levels, exposure to different shocks, degree of exposure, sensitivity (short-term coping), adaptation (survive, cope, recover, learn, transform), which defines how individuals and groups responds to disturbance (FSIN, 2014). While this model primarily applies to disaster risk reduction, other approaches where it can be used include improved livelihoods and climate change adaptation, where these two are integrated into a single framework. This framework puts emphasis on absorptive, adaptive and transformative

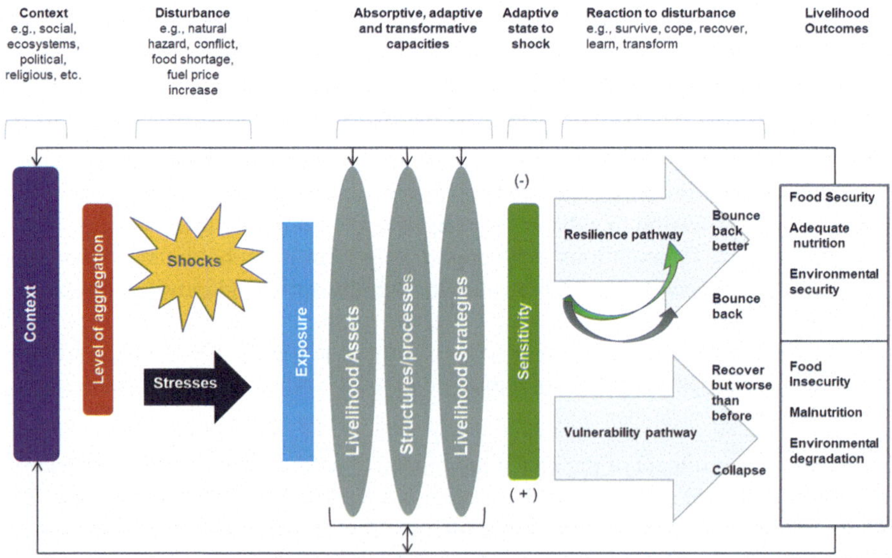

Fig. 5.1 Disaster Resilience Framework. (Source: Food Security Information Network, 2014)

capacities, which entails access to productive assets, livelihood strategies, institutions, integrated with preparedness, prevention, response and recovery activities conveyed to realise well-being outcomes in response to shocks and climate-related stresses (FSIN, 2014). Composite measures (indexes) are created through a combination of variables based on weights, detected through their structure of relationships and correlations (FSIN, 2014).

5.4.3 Economic Modelling of the Relationship Between Household Resilience and Livelihood Assets

We used the random effects probit model for panel data to model the relationship between the resilience capacity index and the set of predictor variables. According to Maddala (1983), the model is structured as Eq. (5.1).

$$y_{it} = \alpha X_{it} + u_{it} \tag{5.1}$$

where Y_{it} are the resilience capacity indices for rounds 1, 2 and 3, X_{it} are the socio-economic, demographic and human capital variables, respectively, that are associated with resilience capacities for individual i at time t, where t = 2019, 2020 and 2021. The model further assumes that $u \sim (0, \partial^2)$ and, therefore, is normally distributed.

Following Eq. (5.1), a description of the response and predictor variables is shown in Table 5.1.

5.5 Results and Discussion

5.5.1 Shocks Faced by Households

Table 5.2 shows the various climatic, economic, idiosyncratic, health-related and other shocks experienced by households across in the seven consortia project areas.

Almost all households (99.7%) have been affected by at least one major shock from 2019 to 2021. Overall, households faced 19 types of shocks, some of which were natural, while others were due to exogenous economic shocks emanating from the unstable macroeconomic environment. Furthermore, households were affected by multiple concurrent shocks, averaging 4.5 shocks across rural households, with some households experiencing more than six shock incidences. The most prevalent climatic shocks in the seven consortia were drought and erratic rainfall. These shocks resulted in low agricultural productivity, deterioration in income and food insecurity. The survey of 2019 was carried out after the heavy flooding resulting from Cyclone Idai and the heat wave that took place in September 2020. Health

Table 5.1 Description of response and predictor variables in Eq. (5.1)

Variables	Description	Variable name	A priori expectation
Dependent Resilience capacity index	1-if the resilience capacity is equal to or greater than the mean; 0-otherwise	res_index	
Predictors			
Human capital			
Household head gender	1 if male; 0 if female	Gender1	+
Age	Household head age in years		+
Education	1-if formally schooled; 0-otherwise	Formalschool	+
Household size	Total number of people aged between 16 and 64 years in household	Hh_mem_total	+
Marital status	1-if married; 0-otherwise	Mstatus	+
Human capital index	Index ranging from 0 to 1	Humancap_ index	+
Financial capital			
Access to financial services	Index ranging from 0 to 1	Finsvs_index	+
Livelihood diversification index	Based on the number of livelihood activities by the household	Livdiv_index	+
Social capital			
Collective action	Index ranging from 0 to 1	Collect_index	+
Linking social capital	Index ranging from 0 to 1	Sclink_index	+
Information on agricultural services index	Index ranging from 0 to 1		
Physical capital			
Productive assets	Index ranging from 0 to 1	Prod_index	+
Community infrastructure	Index ranging from 0 to 1	Inf_index	+
Natural capital			
Land area	Total area owned in hectares	Aa103	+
Livestock assets	Indexed to indicate range and type of livestock owned	lvstk_index	+
Other			
Round	T = 1, 2, 3	Round	
Intensity of participation (high)	1 = if high level participation; 0-otherwise	Highintensity	+
Intensity of participation (medium)	1 = if medium, 0-otherwise	Mediumintensity	+

shocks (illness and large medical expenses) were the major idiosyncratic shocks (11.5%), affecting rural households in the project areas. Sharp food price increase was the most reported market shock reported by 72.1% percent of households. Shocks related to or affecting both crops and livestock production dominated the households under study, particularly the prevalence of crop pests and diseases, and

Table 5.2 Types of shocks faced by households

Shock	Frequency (%)
Overall shock	**99.7**
Climate shocks	
Excessive rains	73
Variable rains	77.7
Floods	67.5
Drought	76.5
Crop and livestock shocks	
Crop pests and diseases	56.1
Reduced soil productivity	19.0
Crop damage by wildlife	14.0
Livestock death or diseases	22.9
Livestock theft	6.3
Economic shocks	
Sharp food price increase	72.1
Increase in price for inputs	39.9
Low price crop or livestock products	14.6
Idiosyncratic household shocks	
Illness of a member	11.5
Large medical expense	10.2
Death of household member	7.1
Other shocks	
Deforestation	5.7
Veld fires	1.6
Covid restrictions	72.7

this is supported by Paumgarten et al. (2020). In support of the results, Debebe et al. (2013) from a study in Ethiopia, found natural shocks being highest at 47%, though economic shocks at 32% tend to be slightly higher than health shocks at 31%. When considered over the three-year period, the absorptive capacity of households improved in the research study areas. In 2019, for instance, the index had an average value of 29, and increased to 32 in 2020, before reaching 36 in 2021. The mean absorptive indices were statistically different at the 5% level ($F = 218; p < 0.05$). For the 3-year period, the average absorptive capacity was 32. This means the households capacity to cope with shocks improved from 2019 to 2021.

5.5.2 *Coping Strategies Implemented by Households*

Using a standardised set of questions, households in the seven consortia were asked to indicate the coping mechanisms associated with each of the 19 shocks identified in the foregoing section. Figure 5.2 shows the different coping strategies which were employed by the households in dealing with shocks.

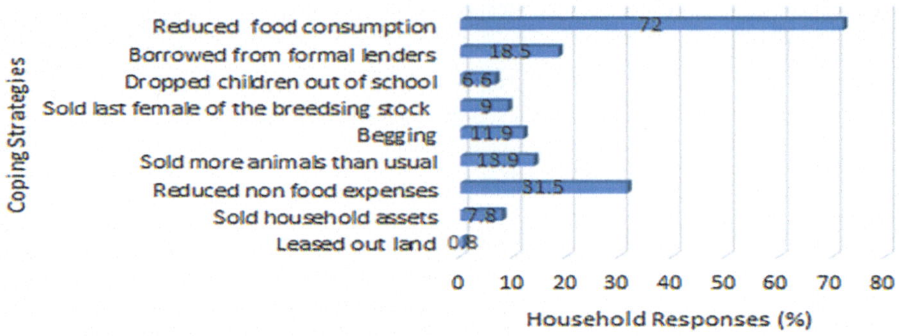

Fig. 5.2 Coping strategies employed by households in dealing with shocks

Table 5.3 Shock preparedness index (SPI)

Year	Mean
2019	18.9
2020	21.8
2021	31.2
Overall	24 ($F = 180, p = 0.00$)

Figure 5.2 shows that reduced food consumption was the principal consumption soothing strategy reported by 72 percent of the affected households. This means the affected households sacrificed their food consumption, which suggests welfare loss, besides diversifying food intake and reliance on food relief. The second prime strategy to enhance resilience was reducing household expenditure towards non-food items (31.5%) followed by sale of livestock portfolios (cattle, goat, sheep and poultry) which was reported by 13.9 percent of the households.

Households also borrowed money from formal lenders including banks and micro financial institutions, (18.5%) suggesting the importance of formal risk sharing mechanism. Furthermore, households sold other assets, including the last female of the breeding stock (9%), other household assets (7.8%) and some dropped their children out of school because they could not pay school fees (6.6%). A very low percentage of affected households leased out land (0.8%).

Further, a shock preparedness index was developed to depict the extent to which households were prepared to deal with unexpected exogenous natural or socio-economic shocks. Table 5.3 shows the mean indices by year and results of the one-way analysis of variance (ANOVA). These results are related to ZRBF participants across all consortia.

Across temporal scales, the shock preparedness index increased by 2.9% from the year 2019 to 2020. The index further increased by 9.4% from 2020 to 2021, which showed that there was enhanced capacity of rural households to withstand the deleterious effect of exogenous natural and economy-based shocks over the study period.

Table 5.4 Association between age of household head and SPI

Pearson correlation coefficient	**Age of household head**
Shock preparedness index	−0.016
Significance (p)	0.076

Table 5.5 Association between gender and SPI

Gender of household head	Mean SPI	N	Standard deviation
1. Male	25.2125	4545	32.90542
2. Female	23.2382	8220	31.07943
Total	23.9411	12,764	31.75440
F-value	11.321		
P-value	0.001**		

5.5.3 *Relationship Between Household Characteristics and Shock Preparedness*

There was a negative relationship between the shock preparedness index and the respondent's age of the household head ($p = 0.07 < 0.1$) (Table 5.4). However, this association was significant at the 10% level. Therefore, young farmers were more vulnerable to shocks since most do not own the strategic assets necessary for recovery. It is assumed that older farmers are more resilient since the likelihood of accumulating strategic assets over time is higher than that of younger farmers.

On the other hand, male-headed households had a higher SPI at 25.2 compared to female-headed households at 23.2 ($p < 0.05$) (Table 5.5). This observation buttresses the notion that male headed households have access to assets such as land and are favoured by the predominantly patriarchal system to easily get access to finance, whereas female headed households are vulnerable with limited access to household assets.

This result is also a reflection of a cultural system that biases resource ownership towards men than women.

5.6 Training Received from ZRBF

Respondents were asked to indicate whether they had participated in ZRBF activities in the years 2019, 2020 and 2021. In this regard, 44% of the sampled farmers participated in at least one activity that was supported by the ZRBF. Table 5.6 shows the types of training that were received across the participating districts as well as the proportion of households that received training.

Table 5.6 Types of training received by farmers

Type of training	Frequency
Quality certified seeds	58.9
Community seed banks	16.6
Adapted suitable improved varieties	53.4
Small grains	65.4
Crop rotation	63.7
Integrated pest management	44.8
Improved shelter	73.6
Vaccinations	46.6
Dipping	69.3
Home-made feeds	36.5
Fodder preservation	29.0
Artificial insemination	17.0
Access to markets	28.6

Most households received training in improved shelter, dipping, crop rotations and small grain production and were, therefore, expected to have a broader understanding of crop and livestock systems. However, specialised activities such as artificial insemination and establishment of community seed banks had been accessed by a few farmers as only 20% of the households had received training in these areas.

5.6.1 Household Characteristics

The mean and significance values of the response and predictor variables included in the regression equations are shown in Table 5.7.

Generally, most of the key descriptor variables increased from waves 1 to 3. The resilience capacity index increased by 20% from wave 1 to wave 2, before rising by a further 8% in wave 3. This index was re-coded as a dummy taking values of 0 if less than 0.48 (mean) and taking values of 1 if more than 0.48. Except for gender, other human capital variables were statistically significant across the three waves at the 5% or 10%. Similarly, financial, social and physical capital predictor variables were statistically different at the 5%. However, the intensity of participation in ZRBF projects across the three waves was not significantly different ($p > 0.05$).

Significant improvements were experienced for social capital across the three waves of household surveys (Fig. 5.3). This implies that social networks (linking social capital) played a more important role in asset accumulation over the 3-year period. On the other hand, and according to interviewed households, no gains were experienced in physical capital (community infrastructure and basic services).

Table 5.7 Measures of central tendency by wave

Variable	Obs	Wave 1	Wave 2	Wave 3	Sig.
Dependent Resilience capacity index	12,870	37.8	41.8	45.5	0.00**
Predictors					
Human capital					
Gender	12,870	0.65	0.65	0.65	0.524
Age	12,592	47.1	48	48.1	0.000**
Education	12,798	0.87	0.9	0.89	0.000**
Household size	8672	–	5.89	5.8	0.072**
Marital status	12,870	0.7	0.73	0.71	0.000**
Human capital index	12,867	56.5	58.3	58.1	0.000**
Financial capital					
Access to financial services	12,870	69.6	79.6	74.8	0.000**
Livelihood diversification index	12,870	8.4	8.2	8.9	0.000**
Social capital					
Collective action	12,870	68	64	67	0.000**
Linking social capital	12,870	26.7	37.2	42.0	0.000**
Physical capital					
Productive assets	12,870	32.9	33.3	35.4	0.000**
Community infrastructure	12,870	65.3	65.9	68.9	0.000**
Natural capital					
Land area	12,870	3.57	3.61	3.01	0.007**
Livestock assets	12,870	63.4	65.1	66.4	0.000**
Other					
Round					
Intensity of participation (high)	12,870	0.5	0.5	0.5	0.93
Intensity of participation (medium)	12,870	0.49	0.49	0.49	0.93

5.6.2 *Factors Determining Household Resilience*

The random effects probit model was used to determine the relationship between household resilience and a set of predictors reflecting the household's social, economic and livelihood assets. Compared to a null, the model was statistically significant at the 5% level ($\chi^2 = 3542$, $p < 0.05$). Table 5.8 shows the random effects probit model results on factors affecting household resilience.

Table 5.9 shows the marginal effects associated with the main predictors included in the model.

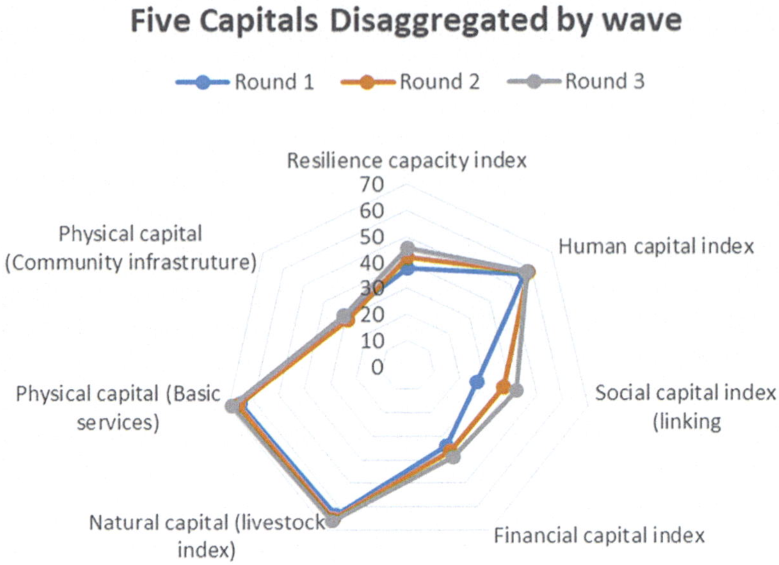

Fig. 5.3 Five capitals disaggregated by round

5.6.3 *Human Capital*

In terms of human capital, household size and age were statistically significant ($p < 0.05$ and $p < 0.1$, respectively). A unit increase in the number of members in a household was associated with a 0.8% increase in the resilience capacity of the household, holding other factors constant. Similarly, a higher human capital index had a positive association with household resilience ($p < 0.05$). Islam and Walkerden (2022) showed that households relied on human capital so much for disaster resilience in Bangladesh. Families with more members were able to recover sooner after a disaster. However, Islam and Walkerden (2022) qualified that the number of working members in the family affects disaster resilience and recovery. This agrees with Jabeen et al. (2010) who found that households with more than one member earning an income did better during recovery after a disaster than those dependent on only one earner. So, in the case of rural households, the more the household members actively provide labour, the higher the household resilience.

Age is an important factor in disaster resilience and recovery. It can be said that young household heads are more vulnerable to exogenous shocks. On the other hand, more life experiences can help households to cope with disaster, although in some cases it might be argued that extreme old age can increase the likelihood of harm during and following disasters (Islam & Walkerden, 2022). The contribution of health and education to productivity are quantified using the Human Capital Index (HCI) (Al-Maruf, 2017). Education plays a pivotal role in human resource development (Rao, 1966) since it enhances human capital (Al-Maruf, 2017). Educated household members are well placed in understanding disaster warning

Table 5.8 Random effects probit model results on factors affecting household resilience

GEE population-averaged model			Number of obs =		6, 650	
Group variable:		round	Number of groups =		2	
Link:		probit	Obs per group:			
Family:		binomial	min =		3297	
Correlation:		exchangeable	avg =		3,325.0	
			max =		3,353	
			Wald chi2 (18) =		3644.92	
Scale parameter:		1	Prob > chi2 =		0.0000	
resindex3.	Coef.	Std. Err.	z	P> |z|	[95% Conf. Interval]	
gender1	.0676003	.0471576	1.43	0.152	–.0248268	.1600274
formalschool	–.1259036	.0835777	–1.51	0.132	–.289713	.0379057
mstatus	.0011772	.0517504	0.02	0.982	–.1002518	.1026062
highintensity	.088227	.0425573	2.07	0.038	.0048162	.1716377
aa103	–.0024668	.001418	–1.74	0.082	–.0052461	.0003125
hh_mem_total	.0443205	.0096394	4.60	0.000	.0254276	.0632133
e601a	.0009675	.0018242	0.53	0.596	–.0026078	.0045427
humancap_index	.0086876	.0011261	7.72	0.000	.0064805	.0108946
livediv_index	.102057	.0060016	17.00	0.000	.090294	.1138199
sclink_index	.0158016	.0005565	28.39	0.000	.0147108	.0168924
collect_index	.007781	.0004623	16.83	0.000	.0068748	.0086871
gender_index	.0030293	.000538	5.63	0.000	.0019748	.0040838
prod_index	.047369	.0021625	21.91	0.000	.0431306	.0516073
lvstk_index	.0099409	.0017093	5.82	0.000	.0065906	.0132911
finsvs_index	.006045	.000444	13.61	0.000	.0051748	.0069153
inf_index	.0058474	.0011774	4.97	0.000	.0035397	.0081552
svs_index	.0062415	.0008365	7.46	0.000	.004602	.007881
round	.2166483	.0080486	26.92	0.000	.2008733	.2324232
_cons	–6.178677	.1913582	–32.29	0.000	–6.553732	–5.803622

signals and have an enhanced ability to take precautionary measures in preparing themselves for disasters by taking actions such as storing food and saving money, which minimise disaster vulnerability and increase resilience (Islam & Walkerden, 2022).

5.6.4 Social Capital

Both collective action and linking social capital variables were positively associated with household resilience ($p < 0.05$). This result reinforces the notion that social capital is indeed one of the critical components in achieving resilience, as households can exploit community-level relationships to recover from unexpected shocks. Ahsan and Warner (2014) found that social capital is influential in adaptive capacity,

Table 5.9 Marginal effects for the random effects probit model

	dy/ dx	Delta-method Std. Err.	z	$P > |z|$	[95% Conf. Interval]	
geoder1	.012844	.0089551	1.43	0.151	−.0047076	.0303955
formalschool	−.0239215	.0158717	−1.51	0.132	−.0550295	.0071864
mstatus	.0002237	.0098325	0.02	0.982	−.0190477	.019495
highintenstity	.016763	.0080777	2.08	0.038	.0009311	.0325949
aa103	−.0004687	.0002692	−1.74	0.082	−.0009963	.000059
hh_mem_total	.0084208	.0018222	4.62	0.000	.0048494	.0119923
e601a	.0001838	.0003465	0.53	0.596	−.0004955	.0008631
humancap_index	.0016506	.0002109	7.33	0.000	.0012372	.002064
livediv_index	.0193907	.0010624	18.25	0.000	.0173085	.0214729
sclink_index	.0030023	.0000841	35.71	0.000	.0028375	.0031671
collect_index	.0014784	.0000819	18.05	0.000	.0013179	.0016389
gender_index	.0005756	.0001015	5.67	0.000	.0003767	.0007745
prod_index	.009	.0003629	24.80	0.000	.0082888	.0097113
lvstk_index	.0018888	.0003223	5.86	0.000	.0012571	.0025204
finsvs_index	.0011485	.0000807	14.24	0.000	.0009904	.0013067
inf_index	.001111	.0002224	4.99	0.000	.000675	.001547
svs_index	.0011859	.0001568	7.56	0.000	.0008786	.0014932
round	.0411629	.0017342	23.74	0.000	.037764	.0445618

sensitivity and exposure to disasters. A study by Masud-All-Kamal and Monirul Hassan (2018) showed that affected communities in Bangladesh rely on social capital through bonding and bridging networks for food and comfort, though over time this capital becomes less profound due to limited financial and physical capital. Another study by Uddin et al. (2020) concluded that community resilience attributes function interactively rather than independently and, therefore, it is essential to understand processes that anchor institutional structures, relations and networks.

5.6.5 *Natural Capital*

The marginal effect associated with livestock assets was 0.18% and this indicates that as this index rises, households' resilience also increases, ceteris paribus. This finding corroborates extensive literature that shows the centrality of livestock as a means of social and economic sustenance for rural households. Studies have shown that this type of asset is vital in the early recovery phase of a disaster (Islam & Walkerden, 2022). Some studies have found out that livestock assets aid in providing buffering capacity to households through smoothing external shocks from weather (Hänke & Barkmann, 2017). Livestock assets thus build resilience to climatic risks (see Acosta, 2021; Alary et al., 2011; Ellis & Mdoe, 2003; Hänke &

Barkmann, 2017). Acosta et al. (2021), Speranza (2010), Fafchamps et al. (1998) and Devereux (1993) indicated that to smoothen the consumption shock as a result of severe drought, livestock holders are compelled to first sell small animals, such as poultry, pigs, sheep and goats; major livestock assets, such as cows, would only be sold when the drought causes more drastic negative shocks. However, Acosta et al. (2021) argues that the buffering potential of livestock varies by income quintile, length of the climate shock, animal species and region.

5.6.6 *Financial Capital*

The importance of income diversification at the household level was confirmed by the positive statistical association between the livelihood diversification index and household resilience ($p < 0.05$). In a context characterised by climate change, portfolio diversification supports the reduction of financial risks that emanate from the rainfed nature of smallholder agricultural activities, the uncertainty of market prices and institutional uncertainties. According to Islam and Walkerden (2022), such assets play a critical role in disaster long-term recovery. To cope with shocks, households primarily relied on savings held at home. Gash and Grey (2016) indicated that the frequently used coping mechanisms when households faced shocks were as follows: selling small livestock ranked first, followed by borrowing from a savings group, reducing food consumption, using savings, borrowing from family and friends and selling grain. In rural areas, purchasing on credit is usually not a preferred method for coping with any of the shocks, although credit from informal sources is common (Gash & Grey, 2016).

5.6.7 *Physical Capital*

Access to essential services and community infrastructure were positively associated with household resilience. The study also noted that the marginal effects of 0.1% reveal that both components are equally crucial in the attainment of household resilience. Islam and Walkerden (2022) concluded in their study that physical capital is an important asset for resilience. Housing quality as part of physical capital is a key indicator of whether a community can withstand cyclones and storm surges. Firm housing structures play a rudimentary role in protecting households and their livelihood assets. Household assets that are used within the homestead for domestic purposes (e.g. furniture) and livelihood assets are basic to households' well-being and contribute to household resilience, especially during natural disasters (Paul, 2009).

5.7 Conclusions and Recommendations

Results suggested that asset building is used as a coping strategy by individuals, families and communities to ensure their socio-economic well-being when they face natural and socio-economic shocks. The assets are both tangible, that is, natural, physical and financial; and intangible, that is, social and human. The assets help households mitigate shock, minimise the severity of trade-offs and avoid long-term adversive developmental outcomes.

Overall, according to the resilience pathway framework, strengthened resilience capacity drives better household responses to shocks and stresses, which leads to better well-being outcomes. The conceptual model posits that increased programme participation drives this pathway, for example, programme participation leads to improvements in resilience capacity, which leads to better responses by households, which in turn helps them maintain or improve their food security or other well-being outcome in the face of shocks or stresses. How do livelihood assets fit into this framework?—for a start they influence programme participation—some households have too few assets to benefit from interventions. Livelihood assets are an important building block for a resilience capacity, but resilience capacity also depends on how these assets are used to absorb or adapt to shocks and stressors. Some assets such as education increase the skills or ability to learn new technologies or access relevant information for shock/risk informed decision-making. Livestock assets can be used to respond to shocks (e.g. sell goats to purchase food). They are also a form of self-insurance to deal with other unfortunate events such as a sudden medical expense or funeral expenses. Livestock assets can also be sold to fund new livelihood activities.

As an insurance against shocks, households should employ a wide range of coping mechanisms to mitigate the detrimental effects of natural and socio-economic shocks. Some of the common ex-ante coping strategies that the households can employ include precautionary savings, which would help even up consumption and diversification of income-generating activities. On the other hand, examples of the ex-post coping strategies include selling productive assets during times of disasters and using formal or informal safety nets. In rural areas, livestock as a natural asset enables vulnerable households to cope with income-reducing shocks. However, livestock-oriented resilience interventions should carefully consider the relationships among species, type of shock and socio-economic groups because the buffering capacity of different animal species varies depending on the length of the shock (Acosta et al., 2021).

Informal safety nets that could be used include provision of credit facilities. There is a need to design financial products for the rural areas with attributes that matter most for coping with shocks: timeliness of pay-out and availability of the product (Gash & Grey, 2016). However, for a very food insecure population, design and offer of the financial products should have an objective biased towards ensuring that the vulnerable population consumes more and better food to improve resiliency. For the very vulnerable food-insecure contexts, financial service providers need to partner with those offering ultra-poor programmes or other support programmes in order to help the vulnerable segment make effective use of financial services (Gash & Grey, 2016).

There is a need to develop more products and mechanisms that help households increase their savings. Savings, whether in the form of livestock or cash, should be promoted as a coping mechanism in anticipation and response to shocks.

References

Acosta, A., Nicolli, F., & Karfakis, P. (2021). Coping with climate shocks: The complex role of livestock portfolios. *World Development, 146*, 105546.

Ahsan, M. N., & Warner, J. (2014). The socioeconomic vulnerability index: A pragmatic approach for assessing climate change led risks–A case study in the south-western coastal Bangladesh. *International Journal of Disaster Risk Reduction, 8*, 32–49.

Al-Maruf, A. (2017). Enhancing disaster resilience through human capital: Prospects for adaptation to cyclones in coastal Bangladesh. Doctoral dissertation, Universität zu Köln, Germany.

Alary, V., Corniaux, C., & Gautier, D. (2011). Livestock's contribution to poverty alleviation: How to measure it? *World Development, 39*(9), 1638–1648.

Asmamaw, M., Mereta, S., & Ambelu, A. (2019). Exploring households' resilience to climate change-induced shocks using climate resilience index in Dinki watershed, central highlands of Ethiopia. *PLoS One, 14*, e0219393.

Bebbington, A. (1999). Capitals and capabilities: A framework for Analysing peasant viability. *Rural Livelihoods and Poverty. World Development, 27*(12), 2021–2044.

CARE. (2009). *Climate vulnerability and capacity analysis handbook*. CARE. https://careclimatechange.org/wp-content/uploads/2014/12/CVCA_EN.pdf

CESVI. (2019). *Annual Report*. CESVI.

Cornelius, D., & Kiiza, W. C. (2021). COVID-19 pandemic and the informal sector in Zimbabwe. *African Journal of Social Work, 11*(4), 201–208.

da Costa, M. D. J., Lopes, M., Ximenes, A., Ferreira, A. D. R., Spyckerelle, L., Williams, R., Nesbitt, H., & Erskine, W. (2013). Household food insecurity in Timor-Leste. *Food Security, 5*(1), 83–94.

Debebe, Z. Y., Mebratie, A., Sparrow, R., Abebaw, D., Dekker, M., Alemu, G. & Bedi, A. S. (2013). *Coping with shocks in rural Ethiopia*. ASC Working Paper Series (110).

Dercon, S., Hoddinott, J., & Woldehanna, T. (2005). Shocks and consumption in 15 Ethiopian villages, 1999 2004. *Journal of African Economies, 14*(4), 559–585.

Devereux, S. (1993). Goats before ploughs: Dilemmas of household response sequencing during food shortages. *IDS Bulletin, 24*(4), 52–59.

DFiD. (2012). *Defining Disaster Resilience: A DFID Approach Paper*. Available at http://www.fsnnetwork.org/sites/default/files/dfid_defining_disaster_resilience.pdf

Eichsteller, M., Njagi, T., & Nyukuri, E. (2022). The role of agriculture in poverty escapes in Kenya—Developing a capabilities approach in the context of climate change. *World Development, 149*, 105705.

Ellis, F., & Biggs, F. (2001). Evolving themes in rural development 1950s–2000s. *Development Policy Review, 19*(4), 437–448.

Ellis, F., & Mdoe, N. (2003). Livelihoods and rural poverty reduction in Tanzania. *World Development, 31*(8), 1367–1384.

Fafchamps, M., Udry, C., & Czukas, K. (1998). Drought and saving in West Africa: Are livestock a buffer stock? *Journal of Development Economics, 55*(2), 273–305.

FAO, IFAD, UNICEF & WFP. (2018). *The state of food security and nutrition in the world 2018*. Building Climate Resilience for Food Security and Nutrition.

Food Security Information Network (FSIN). (2014, November). A common analytical model for resilience measurement. In *Resilience measurement technical working group, technical serion no 2*. WFP.

Ford Foundation. (2004). *Building assets to reduce poverty and injustice*. Ford Foundation.

Franklin, F. T. (2019). The politics of resilience and sustainable development: Pragmatic reflections on vulnerability from environmental, economic and community perspectives. *Research and Science Today, 2*, 125–132.

Gash, M., & Grey, B. (2016). *The role of Financial Services in Building Household Resilience in Burkina Faso*. Consultative Group to Assist the Poor.

Hänke, H., & Barkmann, J. (2017). Insurance function of livestock, farmers coping capacity with crop failure in southwestern Madagascar. *World Development, 96*, 264–275.

Hendriks, S. L. (2015). The food security continuum: A novel tool for understanding food insecurity as a range of experiences. *Food Security, 7*(3), 609–619.

Islam, R., & Walkerden, G. (2022). Livelihood assets, mutual support and disaster resilience in coastal Bangladesh. *International Journal of Disaster Risk Reduction, 78*, 103148.

Jabeen, H. (2015). Adapting the assets of urban low-income households with climate extremes: Experience from Dhaka. *International Journal of Urban Sustain Development, 7*(1), 72–88.

Jabeen, H., Johnson, C., & Allen, A. (2010). Built-in resilience: Learning from grassroots coping strategies for climate variability. *Environment and Urbanization, 22*(2), 415–431.

Jensen, N., & Barrett, C. (2017). Agricultural index insurance for development. *Applied Economic Perspectives and Policy, 39*(2), 199–219.

Kebede, T., Haji, J., Legesse, B., & Mammo, G. (2016). Econometric analysis of rural households' resilience to food insecurity in west Shoa, Ethiopia. *Journal of Food Security, 4*, 58–67.

Maddala, G. S. (1983). *Limited-dependent and qualitative variables in economics*. Cambridge University Press.

Manlosa, A. O., Hanspach, J., Schultner, J., Dorresteijn, I., & Fischer, J. (2019). Livelihood strategies, capital assets and food security in rural Southwest Ethiopia. *Food Security, 11*(1), 167–181.

Manyumwa, H. D., Siziba, S., Unganai, L., Mapfumo, P., & Mtambanengwe, F. (2018). The impacts of community-based cash management tools on smallholder rural farmers' access to livelihood assets. *African Journal of Agricultural and Resource Economics, 13*(2), 182–192.

Marzi, S., Mysiak, J., Essenfelder, A. H., Amadio, M., Giove, S., & Fekete, A. (2019). Constructing a comprehensive disaster resilience index: The case of Italy. *PLoS One, 14*(9), e0221585.

Masud-All-Kamal, M., & Monirul Hassan, S. M. (2018). The link between social capital and disaster recovery: Evidence from coastal communities in Bangladesh. *Natural Hazards, 93*, 1547–1564.

Moser, C. (2006). *Asset-based approaches to poverty reduction in a globalized context: An introduction to asset accumulation policy and summary of workshop findings*. Global Economy and Development.

Mugandani, R., Wuta, W., Makarau, A., & Chipindu, B. (2012). Re-classification of agro-ecological regions of Zimbabwe in conformity with climate variability and change. *African Crop Science Journal, 20*(2), 361–369.

Mwadzingeni, L., Mugandani, R., & Mafongoya, P. L. (2022). Socio-demographic, institutional and governance factors influencing adaptive capacity of smallholder irrigators in Zimbabwe. *PLoS One, 17*(8), e0273648.

Ofoegbu, C., Chirwa, P., Francis, J., & Babalola, F. (2016). Assessing forest-based rural communities' adaptive capacity and coping strategies for climate variability and change: The case of Vhembe district in South Africa. *Environmental Development, 18*, 36–51. https://doi.org/10.1016/j.envdev.2016.03.001

Ostrom, E., & Ahn, T. (2007). The meaning of social capital and its link to collective action. In *Handbook of social capital: The Troika of sociology, political science and economics*. https://doi.org/10.2139/ssrn.1304823

Paul, B. K. (2009). Why relatively fewer people died? The case of Bangladesh's cyclone Sidr. *Natural Hazards, 50*(2), 289–304.

Paumgarten, F., Locatelli, B., Witkowski, E. T., & Vogel, C. (2020). Prepare for the unanticipated: Portfolios of coping strategies of rural households facing diverse shocks. *Journal of Rural Studies, 80*, 91–100.

Prowse, M., & Scott, L. (2008). Assets and adaptation: An emerging debate. *Institute of Development Studies Bulletin, 39*(4), 42–52.

Rahim, S., Saeed, D., Rasool, G. A., & Saeed, G. (2011). Factors influencing household food security status. *Food and Nutrition Sciences*, 1–4.

Rao, V. K. R. V. (1966). *Education and human resource development*. Allied Publisher.

Sherraden, M. (1991). *Assets and the poor: A new American welfare policy*. M.E. Sharpe.

Speranza, C. (2010). Drought coping and adaptation strategies: Understanding adaptations to climate change in agro-pastoral livestock production in Makueni District, Kenya. *European Journal of Development Research, 22*(5), 623–642.

Uddin, M. S., Haque, C. E., & Walker, D. (2020). Community resilience to cyclone and storm surge disasters: Evidence from coastal communities of Bangladesh. *Journal of Environmental Management, 264*, 110457.

UNDP. (2020). ZRBF resilience knowledge hub. In M. Langworthy, K. Fox, S. Martin, L. Woodson, T. Aziz, R. Al-haddad, E. Cuellar, C. Hein, & J. Stack (Eds.), *ZRBF Outcome Monitoring Survey: Round two program learning report*.

WFP. (2011). WFP policy on disaster risk reduction and management. In *Building food security and resilience*. WFP.

World Bank. (2013). *World development report 2014: Risk and opportunity—Managing risk for development*. World Bank.

World Bank. (2016). *Investing in urban resilience: Protecting and promoting development in a changing world*. World Bank.

Zhang, H., Zhao, Y., & Pederson, J. (2019). Capital assets framework for analysing household vulnerability during disaster. *Disasters, 44*(4), 687–707.

ZRBF. (2022). *Zimbabwe resilience knowledge hub presentation*. Website: https://www.resilienceknowledgecentre.co.zw/sites/default/files/2020-09/ZRBF_Resilience%20Storytelling%20Toolkit%20FINAL.pdf

Chapter 6
Environmental and Socio-economic Shocks on Smallholder Farmers: Can Resilience Be Achieved Without Livestock in Zimbabwe?

Clarice Princess Mudzengi, Everson Dahwa, Michael Ticharwa Mubvuma, Lazarus Chapungu, and Xavier Poshiwa

6.1 Introduction

Smallholder farmers are one of the groups highly vulnerable to environmental and socio-economic shocks (Harvey et al., 2018). For instance, climate change poses a significant challenge by subjecting smallholder farmers to abject poverty (Lipper et al., 2014). Other socio-economic shocks, such as global recession, hyperinflation, diseases and market instability also undermine the capability of smallholder farmers to achieve sustainable development (Harvey et al., 2018). Worldwide, there are more than 475 million smallholder farmers living on less than 2 hectares each (Lowder et al., 2016). Most of these farmers are categorised as poor, experiencing food insecurity and living in extremely precarious conditions (Cohn et al., 2017). Dependence on rain-fed agriculture is the major determinant of their vulnerability to environmental and socio-economic shocks. This is exacerbated by the lack of

C. P. Mudzengi (✉)
Research and Innovation Department, Great Zimbabwe University, Masvingo, Zimbabwe
e-mail: cmudzengi@gzu.ac.zw

E. Dahwa · X. Poshiwa
Department of Livestock, Wildlife and Fisheries, Gary Magadzire School of Agriculture and Engineering, Great Zimbabwe University, Masvingo, Zimbabwe
e-mail: edahwa@gzu.ac.zw; xposhiwa@gzu.ac.zw

M. T. Mubvuma
Department of Soil and Plant Science, Gary Magadzire School of Agriculture and Engineering, Great Zimbabwe University, Masvingo, Zimbabwe
e-mail: mmubvuma@gzu.ac.zw

L. Chapungu
Institute for Corporate Citizenship, University of South Africa, Pretoria, South Africa
e-mail: tchapul@unisa.ac.za

L. Mhlanga et al. (eds.), *Building Resilience to Climate Change in Selected Districts in Zimbabwe*, https://doi.org/10.1007/978-3-031-80529-5_6

access to technical and financial support to engage in climate resilient agriculture (Harvey et al., 2018).

In most developing countries, smallholder farmers consider livestock as a source of wealth and a means to sustain livelihoods in the face of environmental and socio-economic shocks (Herrero et al., 2012). Global demand for livestock products is projected to significantly increase by 2050, mainly because of improvements in the standard of living (Rojas-Downing et al., 2017). For example, of the global kilocalorie and protein consumption, livestock accounts for 17% and 33%, respectively (AL-Zoubi et al., 2009). The livestock sector also supports about one billion livelihoods of the world's poorest population and provides employment for approximately 1.1 billion people (Hurst et al., 2005). By 2050, the meat and milk markets in the continent is projected at 34.8 and 82.6 million tonnes, respectively, an increase of 145 and 155 percent, respectively, over 2005/07 levels (AU-IBAR, 2016). The growing demand for livestock products and its rapid growth in developing countries has been deemed the "livestock revolution" (Thornton, 2010; Wright et al., 2012). Despite the "livestock revolution", communities in developing countries are the ones mostly affected by climate change. For instance, warming in Africa is anticipated to be greater than the global average with an increased mean temperature of 3–4 °C over the next century (Zewdie, 2014). This will worsen the adverse impacts of climate change that Africa is already faced with.

Livestock can be an important resource for building resilience to climate shocks through provision of important products that enhance food security. However, many livestock systems are under continual threat of climate shocks such as droughts, which impact quality and quantity of forage, feed crop, water availability and biodiversity. An animal's failure to acquire these resources results in the loss of production and increased disease incidence. In the context of a changing climate, it is critical that we clearly define the role of livestock in building resilience against environmental and socio-economic shocks. For instance, previous frameworks have highlighted the importance of wildlife as insurance against drought (Poshiwa et al., 2013). However, for livestock it is not clear. This raises an important question, which has not been thoroughly explored in the current global literature: Can smallholder farmers in developing countries build resilience to climate shocks through livestock production?

Adverse impacts of climate change–related shocks, such as droughts, livestock diseases and loss on rural livelihoods of Southern Africa, are irrefutable (Rojas-Downing et al., 2017; UNDP, 2020a, b). The Intergovernmental Panel on Climate Change (IPCC) projects scarcity of ecosystem goods and services and increased frequency and intensity of loss event due to climate change, and this is expected to continue in future (IPCC, 2014; Field et al., 2014). It is thus expected that sustainability of most rural livelihoods in semi-arid areas, which are based on rain-fed agriculture, including cropping and livestock husbandry, will be affected. For instance, semi-arid areas are characterised by extensive livestock production systems that are dependent on rangelands. Due to the projected increases in temperature and decreases in precipitation, rangeland productivity will be reduced, leading to decreases in cattle production and productivity, and consequently rural livelihoods. However, few studies have determined the extent to which livestock production can help farmers adapt to climate change. Focus of these studies is mainly on reducing risks due to climate change. Therefore, there is a need for evidence to inform decision-making, including prioritisation and upscaling of emergency disaster responses for communities to build long-term adaptive capacity. In this study, we assert that without livestock, smallholder farming communities may not achieve resilience against climate shocks.

Whilst the Outcome Monitoring Survey (OMS) report by the Zimbabwe Resilience Building Fund (ZRBF) provides a very important overview and assessment of the contribution of livestock production at the household level in coping with environmental and socio-economic shocks, this chapter provides a more granular assessment to ascertain the significance of livestock in shock impact mitigation and adaptation in three consortia, the Enhancing Community Resilience and Sustainability (ECRAS) in Chiredzi and Mwenezi, the Enhancing Community Resilience and Inclusive Market Systems (ECRIMS) in Mberengwa and the Programme for Growth and Resilience (PROGRESS) in Beitbridge and Nyanga districts. The expected outcomes are improved comprehension of the importance of livestock production in building resilience against climate shocks and more appropriate coping mechanisms for livestock-based livelihoods for environmental and socio-economic shocks.

6.1.1 Environmental and Socio-economic Shocks on Smallholder Farmers: An Overview

Worldwide, food value chains are dominated by smallholder farmers who constitute approximately 85 percent of the farms in the world (FAO, 2016). However, in most developing countries, smallholder farmers are faced with several challenges, including the occurrence of negative events that affect all facets of life, commonly referred to as shocks (Ngenoh et al., 2018). Extreme natural events such as landslides, droughts, floods and storms are likely to precipitate shocks. Market-related events such as price fluctuations, shortages of product supplies and input volatilities can also introduce shocks amongst smallholder farmers (Ngenoh et al., 2018). The shocks, regardless of their origin, affect production systems, local economies and food markets. This plunges smallholder farmers into the poverty cycle as all their means of production and livelihoods are persistently hit by environmental and socio-economic shocks.

Environmental shocks, fuelled largely by climate change constitute one of the enormous challenges encountered by smallholder farmers worldwide. The environmental challenges then exacerbate the proliferation of socio-economic shocks threatening smallholder farmers' livelihoods and increase poverty levels (Kom et al., 2020). This has long-term impacts on the livelihoods and development of smallholder farmers (Gloede et al., 2013). Table 6.1 shows the types of shocks and their effects on farmers.

Table 6.1 Categories of shocks and their effects. (Source: Ngenoh et al., 2018)

Type of shock	Examples	Effects
Production shocks		
Weather-related	Drought, water shortage and rainfall or temperature variability	Lower yields, loss of productive assets or income
Natural disasters	Flood, storm, unusually heavy rain, landslides/ erosion	Lower yields, loss of productive assets or income
Biological	Pest, diseases, contamination	Lower yields, loss of income
Marketing shocks		
Prices	Changes in market demand and supply, volatilities, hikes in fuel, input and output prices	Lower produce prices, loss of income, low-quality products

The impacts of shocks on smallholder farmers have previously been extensively studied, for instance in Pakistan (Heltberg & Lund, 2009) and in Swaziland (Mabuza et al., 2016). However, there has not been a thorough review of the decisions to cope with the various shocks. More so, no analysis has been done to ascertain if resilience and dexterity to cope can be achieved without livestock. Such knowledge gaps prompt the need to investigate the role of livestock in the smallholder farmer adaptation and resilience matrix.

6.1.2 Livestock-Based Resilience to Environmental and Socio-economic Shocks

The ZRBF conducted three OMS surveys on resilience strengthening between 2019 and 2021. Findings from round 1 of the OMS surveys indicate that households in most intervention areas have been exposed to a diversity of environmental and socio-economic shocks (UNDP, 2020a, b). Several absorptive, adaptive and transformative coping strategies have been adopted at various levels. At the household level, livestock production contributes about 13.3% of the cash source under the ECRAS consortium whilst contributing 9.7% under the PROGRESS consortium and 10.5% under the ECRIMS. Livestock sales were ranked number 3 out of 16 most important sources of income under the ECRAS consortium, whilst in the PROGRESS and ECRIS consortia, they were ranked sixth and fourth, respectively. Livestock product sales, including provision of draught power, was ranked 14th out of the 16 cash sources. However, under the PROGRESS consortium, livestock product sales was ranked last. In terms of climate smart practices through livestock practices, round 1 report shows that about 86.4% of the households under ECRAS are familiar with climate smart practices, whilst under the PROGRESS and ECRIMS projects, 52.2 and 63.3%, respectively, are familiar. Whilst households with livestock had very important sources of income to use to address other shocks, the report indicates that for a significant number of the households (67.4%), once their livestock is affected by the shocks, they fail to recover.

Round 2 report of the OMS shows that livestock sales increased in their importance as a coping strategy for environmental and socio-economic shocks, from 24.8% in round 1 to 29.1%. Livestock sales are in the top ten coping strategies used by the households. Although the percentage of households indicating they depend on livestock as a cash source decreased from 31.4% in round 1 to 29.6% in round 2, livestock production and sales was the second most important source of cash after casual labour. Thus, a deeper comprehension of the significance of livestock, particularly in building resilience to climate shocks, is necessary.

6.1.3 Household Resilience Capacity Index

The ZRBF defines resilience as "the ability of at-risk individuals, households, communities, and systems to anticipate, cushion, adapt, bounce back better, and move on from the effects of shocks and hazards in a manner that protects livelihoods and recovery gains, and supports sustainable transformation". Resilience Capacity Index is made up of absorptive capacity, adaptive capacity and transformative capacity (Fig. 6.1).

Absorptive capacity is the ability, where possible, to reduce exposure to future shocks and stresses, and for rapid recovery during or after exposure. Improved disaster risk management seeks to strengthen absorptive capacity at the community and household levels, thus aiding them to minimise disaster risk and absorb the impacts. Adaptive capacity includes proactive and informed decisions about alternative livelihood strategies based on changing conditions. Adaptive capacity strategies are aimed at enhancing the flexibility of households and communities to

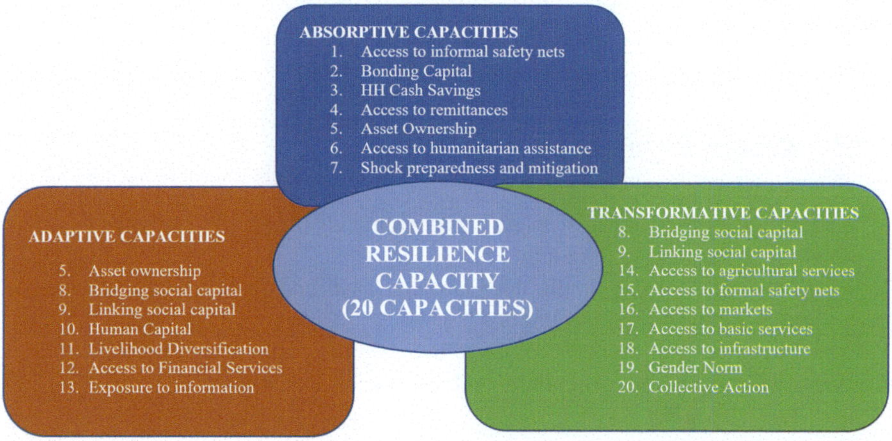

Fig. 6.1 Components of Resilience Index. (Source: ZRBF Second Round Report, 2019)

respond to longer-term social, economic and environmental change. This includes promoting livelihood diversification, assisting in asset accumulation and improving the social and human capital that vulnerable populations get. Of the seven components of the adaptive capacity index computation, "assets ownership" is common for both absorptive and adaptive capacity indices. The six components that were used by ZRBF to compute the adaptive capacity index were bridging social capital score, linking social capital score, human capital, livelihood diversification, access to financial services and exposure to information.

Transformative capacity comprises the governance mechanisms, infrastructure, policies/regulations, community networks and formal and informal social protection mechanisms that make up an enabling environment for systemic change. Improved social capital, livelihood strategies and effective community-level planning to cope with shocks and stresses help to achieve both absorptive and adaptive capacities within a relatively shorter term. However, transformative capacity is dependent on system-level change which takes more time to achieve. The transformative capacity index was calculated using nine components, where bridging and linking social capital are common among both the adaptive and transformative capacity indices. Figure 6.1 shows the components that were used by ZRBF to compute the transformative capacity index.

6.2 Research Purpose

The purpose of this research was to determine the significance of livestock production in enhancing smallholder farmers' resilience to environmental and socio-economic shocks in Zimbabwe. The concept of resilience is rapidly gaining recognition as an alternative and strategic response to anticipated impacts of climate change globally. It is considered a conceptual framework that can help understand how communities can adapt to climate change (Wilson, 2012; Bahadur et al., 2013; UNDP, 2021). Therefore, this is the main purpose of this research.

6.2.1 Research Questions

1. What is the contribution of livestock towards increasing shock resilience amongst smallholder farmers in Zimbabwe?
2. To what extent are the smallholder farmers' cash and food sources sustainable without livestock-based livelihoods?
3. What strategies can be put in place to increase the resilience of livestock-based livelihoods against environmental and socio-economic shocks?

6.3 Methodology

6.3.1 Study Site

The study covers districts within the ZRBF's resilience building interventions under the ECRAS, ECRIMS and PROGRESS consortia run by partner organisations. These include Chiredzi, Mwenezi (ECRAS), Mberengwa (ECRIMS) and Beitbridge and Nyanga districts (PROGRESS) in Zimbabwe (Fig. 6.2). The three southernmost districts (Chiredzi, Beitbridge and Mwenezi) lie in the arid to semi-arid regions of Zimbabwe. They are representative of large areas of semi-arid land in Southern Africa where rain-fed agricultural production is the mainstay of the rural population but is increasingly under threat from climate change and related environmental and socio-economic factors.

The three southernmost districts in Zimbabwe (Mwenezi, Chiredzi and Beitbridge) have almost similar climate characteristics, and the impacts of climate change are almost identical.

Mwenezi District is partly located in Natural Region IV and V of the Zimbabwe agro-ecological zonation, at 21.4226° S, 30.7264° E. The area is characterised by hot and dry weather throughout the year. For instance, it receives low and erratic

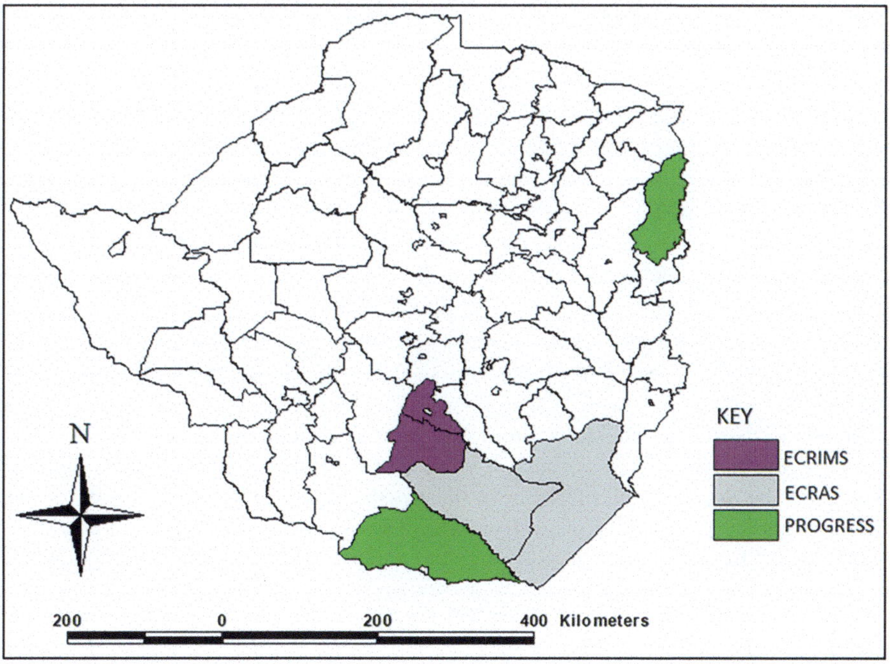

Fig. 6.2 Map of Study area

rainfall with an annual mean of 550 mm (Paenda et al., 2020). Summer temperatures are also very high, exceeding 25 °C, while winter temperatures are low, ranging between 10 and 15 °C, resulting in a high annual range of 10–15 °C (Frischen et al., 2020). Thus, the district is prone to frequent and high intensity meteorological droughts. Most of the communities and households in the district depend on rain-fed crop production and livestock rearing. The livelihood systems are biased towards livestock rearing (specifically goats and cattle) because of the meteorological factors as well as the changing climatic conditions, which have constrained crop production in many ways. The common browse trees in the district include *Colophospermum mopane.*

Chiredzi District is located between 18°55′S and 29°49′E, in Masvingo Province, a region characterised by aridity and uncertain rainfall patterns. It receives average annual rainfall in the range of 450–750 mm in the southern and north-eastern parts of the district, respectively (Jiri et al., 2017). Most of the areas in the district are not suitable for arable agriculture, hence the allocation of most of the land to wildlife conservation. Most households depend on sorghum, which is drought tolerant, requiring minimal rain to grow to maturity. Some parts of the district with red soils are under sugarcane production. There is significant reliance on livestock farming at commercial, small-scale commercial and subsistence levels. The crops that are commonly grown in the area are maize, sorghum, millet, cowpeas, sunflower and cotton. Maize, however, has high failure rate due to low rainfall. Livestock production is mainly affected by a widespread shortage of grazing in the district.

Beitbridge Rural District is in Matabeleland South Province. It comprises 15 wards with 58 villages settled by approximately 23,000 households and 81,588 people (ZimStats, 2012). The district receives low and erratic rainfall, averaging 250–450 mm annually between November and March. The soils are dominantly clay to clay loam, while vegetation is mainly dominated by Miombo, *Burkea africana, Terminalia sericea* and *Colophospermum mopane* tree species. Graminae species dominating the district include the *Eragrostis* spp. and *Digitaria* spp. Most households depend on agro-based livelihoods characterised mainly by livestock production, while crop production is done at a small scale. A diversity of livestock species is kept by most of the households, including cattle, goats, donkeys, pigs and sheep.

Nyanga District lies in Manicaland Province. It covers an area of 589,782 km^2 and is in the north-eastern corner of Manicaland. It shares borders with Mutasa District in the South, Mudzi District in the North, Makoni District in the West and Mozambique in the East. Areas in the district fall under Natural Regions I to V. It is characterised by *Adansonia digitata* and *Acacia* spp. implying that the area generally receives low rainfall (Mushore et al., 2021).

6.3.2 Data Sources and Analysis

The study employed quantitative and qualitative data obtained during the OMS of the ZRBF (UNDP, 2020a, b). The quantitative component of the OMS involved the use of a structured questionnaire designed to collect information for the ten ZRBF top-line indicators and other relevant information required for the descriptive analysis (UNDP, 2020a, b). For the qualitative component, the OMS used three tools to guide data collection, including topical outlines for community level, district level, institutional level focus group discussions (FGDs) and key informant interviews. This approach is mainly important for monitoring resilience programming, which focuses on strengthening capacities, shocks/stresses, coping and livelihood strategies and well-being outcomes (UNDP, 2020a, b).

To improve our understanding of the role of livestock in resilience, we focus on livestock production and sales data generated during the three rounds of OMS for the PROGRESS, ECRIMS and ECRAS consortia. Ranking of the coping mechanisms had already been done in the ZRBF reports. The livestock component was factored out to determine resilience with and without livestock production. Thus, resilience with and without livestock production at the household level was determined and compared to ascertain whether the households can cope without the livestock component of their livelihoods. Data on livestock metrics (numbers, sales) was regressed (linear regression) against the Resilience Index (RI) weighted with drought, livestock theft, low market prices and livestock diseases as covariates (Berghof et al., 2019). The assessment was based on regression coefficient (R^2) and the P-value (probability of accepting the hypothesis at 95% confidence interval). Linear regression analysis was also used to evaluate the extent to which smallholder farmers' cash and food sources are sustainable without livestock-based livelihoods. In this case, cash sources related to livestock sales were regressed against RI to establish the level of relationship between the two parameters. The RI was determined within the ZRBF database as the function of total coping strategies, exposure, sensitivity and farmers' adaptive capacity. Assessment of the existence of strategies that can increase the resilience of livestock-based livelihoods against environmental and socio-economic shocks was done using a non-parametric Mann-Kendall test and Pearson product moment correlation test. Statistical correlation in livestock practices/strategies was assessed using the Kendall's tau, where, values of $> + 0.6$ and $< - 0.4$ were considered to have strong positive and weak relations, respectively (Hennemuth et al., 2013). The MK test was used due to its advantage of being simple, robust and less sensitive to outliers and missing data (Ngongondo et al., 2011). The MK test is less sensitive to abrupt changes in data trends that may be due to non-homogeneity or linearity in dataset (Hennemuth et al., 2013). The MK test is computed as follows:

$$Z_{MK} = \{ \begin{array}{ll} \frac{s-1}{\sqrt{var\, s}} & \text{if } s > 0 \\ 0 & \text{if } s = 0 \\ \frac{s+1}{\sqrt{var(s)}} & \text{if } s < 0 \end{array} \tag{7.1}$$

S is defined as follows:

$$S = \sum_{i-1}^{n-1} \sum_{j=i+1}^{n} sign\left(x_j - x_i\right) \tag{7.2}$$

Where xj and xi are sequential data values (Kendall, 1975; Mann, 1945).

SPSS software version 25 was used to perform the MK test. The null hypothesis was tested at 95% confidence level for all parameters under evaluation.

6.4 Results and Discussion

6.4.1 Resilience Capacity Indices and Resilience Capacity Elements, OMS Rounds 1–3 (Mean Score)

Table 6.2 depicts resilience capacity indices and resilience elements that were determined during the OMS 1, 2 and 3. All the resilience capacities and their elements showed significant changes in all the OMS. Of particular significance was the contribution of livestock towards resilience capacity of farmers. Livestock, which falls under absorptive capacity index, significantly increased ($P < 0.01$) from 63.4 to 65.2 during the period from OMS 1 to OMS 2, whilst it increased significantly ($P < 0.001$) to 66.5 during the period from OMS 2 to OMS 3. Indeed, the upward increase of absorptive capacity index of livestock from round 1 survey to 2 and 3 clearly supports the hypothesis that ZRBF resilience programming strengthens household resilience capacity.

6.4.2 The Contribution of Livestock Towards Shock Resilience Amongst Smallholder Farmers

Generally, results show that when faced with shocks, in all three consortia, farmers could achieve resilience by selling livestock (Table 6.3). However, during 2019, livestock sales could not significantly ($P > 0.05$) help farmers to achieve resilience when farmers were faced with livestock deaths from diseases, low prices or livestock theft. Livestock sales significantly facilitated achievement of resilience when

Table 6.2 Resilience capacity indices and resilience capacity elements, OMS rounds 1–3 (mean score)

	*R*1	*R*2	*R*1 to *R*2	*R*3	*R*2 to *R*3	*R*1 to *R*3
Resilience capacities and their elements (mean, 0–100)						
Absorptive capacity index	29.9	32.2	***	36.0	***	***
ISN	14.8	19.5	***	27.7	***	***
Bonding social capital	47.6	50.0	ns	56.1	*	***
Cash savings	13.3	17.4	**	19.8	*	***
Remittances[a]	11.3	11.9	ns	12.0	ns	ns
Productive assets	33.0	33.3	ns	35.4	***	***
Livestock	63.4	65.2	**	66.5	**	***
Humanitarian assistance	86.8	91.1	**	86.6	***	ns
Shock preparedness	19.0	21.8	*	31.2	***	***
Adaptive capacity Index	31.8	34.8	***	37.6	***	***
Bridging social capital	34.2	37.5	*	40.7	*	**
Human capital	56.6	58.3	**	58.2	ns	*
Livelihood diversification	6.8	8.6	*	7.7	ns	ns
Information exposure	42.5	50.1	***	56.3	***	***
Transformative capacity index	51.6	56.0	***	59.0	***	***
Access to ag services	68.4	76.8	***	83.1	***	***
Formal safety nets (FSN)	69.7	79.6	***	74.8	**	**
Markets	30.4	32.2	ns	38.7	***	***
Basic services	65.4	66.0	ns	68.6	**	***
Infrastructure	30.9	28.9	***	30.8	**	ns
Gender norms	42.3	51.9	***	50.2	ns	***
Resilience capacity index	37.8	41.8	***	45.5	***	***
n	*3440*	*3353*		*3389*		

ns not statistically significant; † $p < 0.1$, * $p < 0.05$, ** $p < 0.01$, *** $p < 0.001$

[a]Remittances was not included in the absorptive capacity index because it had a negative component score

Sources: UNDP, 2020a, b, 2021. ZRBF Outcome Monitoring Surveys [OMS]

Source: ZRBF OMS Report 3

Table 6.3 Livestock sales and Resilience Index under drought, livestock theft, low market prices and livestock diseases

	2019		2020		2021	
Climate and socio-economic shocks	Adjusted R^2	*P*-value	Adjusted R^2	*P*-value	Adjusted R^2	*P*-value
Drought	0.018	0.000*	0.004	0.018*	0.011	0.002*
Livestock theft	−0.005	0.784	0.010	0.112	0.021	0.343
Low livestock prices	0.002	0.236	0.004	0.194	0.013	0.065
Livestock death from diseases	0.003	0.163	0.017	0.004*	0.002	0.163

faced with drought. In 2020, selling livestock significantly ($P < 0.05$) helped farmers to achieve resilience after being exposed to drought or death of livestock due to diseases. Contrarily, livestock sales did not significantly ($P > 0.05$) help farmers to achieve resilience when faced with low prices and livestock theft. During 2021, livestock sales helped farmers in all the three consortia to achieve resilience when farmers were faced with drought but could not help them when they faced low market prices, or livestock theft or death due to livestock diseases. Indeed, it has been reported that increase in livestock sales may result in falling of market prices (Bahta & Myeki, 2021). Thus, farmers are encouraged to form groups or commodity associations that will manage livestock product supply and manage the prices at optimum levels.

In addition to livestock, we also deduced from our results that other non-livestock assets contributed towards resilience of smallholder farmers when faced by different shocks. It is therefore important for farmers to depend on these non-livestock assets to preserve livestock in times of low market prices. The results observed from the three consortia were expected, especially with them being constituted by the dry areas of Chiredzi and Mwenezi, where livestock production makes more economic sense than cropping with regard to climate change and variability. Results from some areas like Inyanga might not have been reflective of the role of livestock sales in building resilience of farmers, as the area is in different agro-ecological regions and falls under a lesser scale of livestock production than the rest of the areas studied. The study results are supported by earlier studies which reported the contribution of livestock sales towards resilience capacity of smallholder farmers (Matlou et al., 2021).

6.4.3 Relationship Between Different Types of Livestock and Resilience Capacity Index of Smallholder Farmers

Tables 6.4a, b and c depict partial regression results from the three surveys conducted in 2019, 2020 and 2021 in the three consortia. In 2019, smallholder farmers who owned donkeys, pigs, sheep and goats were able to significantly ($P < 0.05$) achieve resilience, unlike farmers who were running poultry enterprises only (Tables 6.4a, b, and c). In 2020 also, owning cattle, pigs, sheep and goats and rabbits helped achieve resilience. Similarly, owning cattle, pigs, sheep and goats and poultry was helpful in achieving resilience. Farmers who owned cattle, pigs, sheep and goats and poultry in 2021 were significantly ($P < 0.05$) able to achieve resilience. The results clearly show that livestock has a strong effect on the level of resilience of the farmers. The results suggest that the type and numbers of livestock owned by a household determines the family's resilience (Modal et al., 2023). Indeed, families with large cattle for example in Zimbabwe are more resilient when faced with drought and climate change (Dahwa et al., 2022). Previous scholars have highlighted the impacts of different shocks to resilience of farmers (Mdungela et al.,

Table 6.4a Correlation between different types of livestock and Resilience Index

Control variables			Donkeys	Pigs	Sheep and goats	Cattle	Poultry
Resilience capacity index	Donkeys	Correlation	1.000				
		Significance (2-tailed)	.				
	Pigs	Correlation	0.050	1.000			
		Significance (2-tailed)	0.829	.			
	Sheep and goats	Correlation	0.457	0.350	1.000		
		Significance (2-tailed)	0.037	0.119	.		
	Cattle	Correlation	0.607	0.212	0.330	1.000	
		Significance (2-tailed)	0.003	0.355	0.144	.	
	Poultry	Correlation	0.238	0.088	0.097	0.292	1.000
		Significance (2-tailed)	0.299	0.706	0.677	0.199	.

Table 6.4b Correlation between different types of livestock and Resilience Index

Control variables			Sheep and goats	Cattle	Donkeys	Rabbits	Pigs	Poultry
Resilience capacity index	Sheep and goats	Correlation	1.000					
		Significance (2-tailed)	.					
	Cattle	Correlation	0.493	1.000				
		Significance (2-tailed)	0.000	.				
	Donkeys	Correlation	0.555	0.437	1.000			
		Significance (2-tailed)	0.000	0.000	.			
	Rabbits	Correlation	−0.019	0.023	−0.028	1.000		
		Significance (2-tailed)	0.480	0.377	0.297	.		
	Pigs	Correlation	0.037	0.110	0.028	0.051	1.000	
		Significance (2-tailed)	0.164	0.000	0.285	0.054	.	
	Poultry	Correlation	0.148	0.165	0.097	0.120	0.007	1.000
		Significance (2-tailed)	0.000	0.000	0.000	0.000	0.789	.

2017; UNDP, 2020a, b; UNDP, 2021). The effect of drought shocks on resilience of communities has also been reported in literature and is often coupled with recommendation to increase crop productivity (Mdungela et al., 2017). However, instead of intensifying crop production, we proffer an alternative way of using livestock

Table 6.4c Correlation between different types of livestock and Resilience Index

Control variables			Rabbits	Pigs	Poultry	Sheep and goats	Donkeys	Cattle
Resilience capacity index	Rabbits	Correlation	1.000					
		Significance (2-tailed)	.					
	Pigs	Correlation	0.007	1.000				
		Significance (2-tailed)	0.781	.				
	Poultry	Correlation	0.073	0.067	1.000			
		Significance (2-tailed)	0.005	0.010	.			
	Sheep and goats	Correlation	0.006	0.052	0.186	1.000		
		Significance (2-tailed)	0.826	0.046	0.000	.		
	Donkeys	Correlation	−0.006	0.013	0.072	0.441	1.000	
		Significance (2-tailed)	0.812	0.606	0.006	0.000	.	
	Cattle	Correlation	0.073	0.064	0.158	0.417	0.353	1.000
		Significance (2-tailed)	0.005	0.014	0.000	0.000	0.000	.

assets to improve resilience of communities against shocks such as drought. It is also important to note that low livestock prices may reduce income for the farmers and hence reduce the resilience capacity to withstand shocks (Mandarino et al., 2019).

6.4.4 *To What Extent Are Smallholder Farmers' Cash and Food Sources Sustainable Without Livestock-Based Livelihoods?*

Total cash sources from livestock products sales significantly ($P < 0.05$) contributed towards resilience under drought shocks in all the three consortia, in OMS 1–3 (Table 6.5). However, total cash sources were significantly ($P < 0.05$) helpful in achieving resilience in 2019 when farmers were faced with drought, livestock theft, low market prices and livestock disease shocks. This is contrary to observations in 2020 and 2021 where total cash sources from livestock products sales had no significant contribution ($P > 0.05$) when farmers faced livestock theft, but had a significant effect under drought, low market prices and livestock disease shocks. Results indicate that under drought conditions, total cash sources from livestock products may significantly contribute towards resilience of farmers (Mondal et al., 2023; Bahta, 2022). This is expected, as farmers may reduce their livestock to have sustainable numbers that can be supplemented with available feed resources. In

Table 6.5 Relationship between total cash source from livestock products sales and Resilience Index under drought, livestock theft, low market prices and livestock disease conditions

	2019		2020		2021	
Climate and socio-economic shocks	Adjusted R^2	*P*-value	Adjusted R^2	*P*-value	Adjusted R^2	*P*-value
Drought	0.050	0.000*	0.025	0.000*	0.022	0.000*
Livestock theft	0.034	0.008*	−0.002	0.382	0.015	0.086
Low livestock prices	0.052	0.001*	0.021	0.033*	0.087	0.000*
Livestock death from diseases	0.048	0.000*	0.007	0.049*	0.020	0.001*

*shows significant differences at ($P < 0.05$)

Table 6.6 Livestock ownership, OMS rounds 1–3 (mean number owned)

	R1	R2	R1 to R2	R3	R2 to R3	R1 to R3
Cattle	3.7	4.2	**	4.5	*	***
Draught cattle	1.0	1.3	***	1.5	***	***
Donkeys/mules	0.9	0.8	†	0.8	ns	†
Sheep & goats	5.9	6.3	†	6.7	†	***
Pigs	0.2	0.2	ns	0.2	ns	ns
Poultry	10.5	10.5	ns	12.4	***	***
Rabbits	0.2	0.1	ns	0.1	ns	ns
Livestock value, 000's USD2019	1.5	1.7	*	1.8	**	***
Livestock TLU equivalent	4.2	4.8	***	5.2	**	***
n	*3373*	*3292*		*3324*		

Sources: UNDP 2019, 2020a, b, 2021. ZRBF Outcome Monitoring Surveys (OMS)
ns not statistically significant; † $p < 0.1$; * $p < 0.05$, ** $p < 0.01$, *** $p < 0.001$

Chiredzi, it has been reported that farmers intentionally keep large numbers of livestock, especially cattle as an insurance against drought. Farmers with large number of cattle have been reported to be more resilient to drought shocks (Mondal et al., 2023). Poultry and some other small livestock are also easily disposable, and their consumption increases compared to non-drought years.

6.4.5 Livestock Ownership, OMS Rounds 1–3: Mean Number Owned

Table 6.6 shows livestock ownership in OMS rounds 1–3. Cattle and drought cattle ownership increased significantly ($P < 0.05$) during transition from OMS 1 to 2. However, ownership of other livestock (donkeys, poultry, rabbits, sheep and goats did not vary significantly ($P > 0.05$) during the transition from OMS 1 to 2. Significant ($P < 0.05$) changes in ownership were observed during transitioning from OMS 2 to 3 in cattle, drought cattle and poultry. The significant increase in livestock numbers from round 1 to round 3 was possibly caused by practices such as routine health management, vaccination, castration, deworming/dosing, dipping etc., which are also encouraged to improve livestock production and productivity. Generally, other non-livestock assets also contributed towards farmers` resilience.

6.4.6 *Significance of Good Livestock Practices to Achieve Resilience Against Shocks*

Results show that when faced with shocks, in all three consortia, farmers could achieve resilience by conducting good livestock practices (Table 6.7). Livestock practices under evaluation included keeping of indigenous or improved livestock breeds, construction of appropriate animal shelter and establishment of feeding and watering infrastructure. These livestock practices also included routine health management practices such as vaccination, castration, deworming/dosing and dipping.

In general, the study results encouraged good livestock practices to significantly ($P < 0.05$) improve resilience capacity of farmers through increased livestock production and productivity. For instance, during 2019, under the three consortia, livestock practices significantly ($P < 0.01$) helped farmers to achieve resilience when they were faced with drought and livestock death from diseases. However, good livestock practices did not help farmers to achieve resilience capacity when they were exposed to low prices and livestock theft, unlike when faced with livestock deaths from diseases ($P < 0.05$). Similar results were observed in 2020 and 2021, except that in 2020, farmers were able to significantly ($P < 0.01$) achieve resilience capacity when faced with low livestock prices. Perhaps the depressed prices were managed through supply of better quality of livestock produced under good practices. Elsewhere, good livestock practices have been reported to amortise costs of production and improve market prices (Mandarino et al., 2019).

Findings in this study indicated the significant contribution of good livestock practices on resilience of farmers against all shocks, in the three consortia. The livestock practices under evaluation included improved livestock breeds, improved animal shelter, appropriate water infrastructure and feeds and feeding practices. These results are not surprising, as supplementary feeding, good health

Table 6.7 Relationship between livestock practice and farmers' resilience capacity index when exposed to shocks (drought, livestock theft, low market prices and livestock diseases)

Climate and socio-economic shocks	2019		2020		2020	
	Kendall's tau	*P*-value	Kendall's tau	*P*-value	Kendall's tau	*P*-value
Drought	0.529**	0.000	0.537**	0.000	0.565**	0.001
Livestock theft	0.611**	0.000	0.369**	0.000	0.293	0.362
Low livestock prices	0.341**	0.005	0.017	−0.023	0.423	0.000
Livestock death from diseases	0.751*	0.021	0.464**	0.000	0.452**	0.000

* shows significant differences at ($P < 0.05$)
** shows significant differences at ($P < 0.01$)

management and improved shelter among other livestock practices have been shown earlier to increase livestock production and productivity (Sekhonyana & Fulpagare, 2015; Serrato-Corona et al., 2010). In the tropics, animals require approximately 7% crude protein for adequate neutral detergent fibre use from low-quality basal forage diets (Lazzarini et al., 2009). However, in most dryland areas, such as Beitbridge under PROGRESS, feed shortages are often experienced during the prolonged dry seasons characteristic of these areas. The main supplementary feed resources such as cereal crop residues also have low nutritive value (Ngongoni et al., 2007). Home based feeds using locally available resources, pen fattening and fodder conservation are some of the practices reported in literature, that, congruent to results from this study, can also improve livestock feed nutrition. Thus, any practices that improve both quantity and quality of feeds can increase feed intake, degradability and digestibility. Livestock mortalities related to poor nutrition are often reported in areas such as Chiredzi and Mwenezi, particularly during droughts (Murungweni et al., 2012). Similarly, livestock mortalities due to diseases can also be reduced through routine health management practices such as dosing, dipping and regular vaccinations. Indigenous livestock breeds such as the Mashona cattle are also hardy and less susceptible to diseases than exported breeds.

Results show a significant relationship between livestock practice and farmers' resilience capacity after exposure to drought, livestock theft, low livestock prices and livestock death from diseases (Table 6.7).

6.4.7 What Strategies Can Be Employed to Increase the Resilience of Livestock-Based Livelihoods Against Environmental and Socio-economic Shocks?

Based on the findings from this study and the existing literature spanning the geographic scope like those of the studied areas, several strategies can be put in place to increase the resilience of livestock-based livelihoods against environmental and socio-economic shocks. Table 6.8 shows some of the strategies.

Failure or success of these strategies is dependent on other farmer support systems such as government subsidies, price controls, farmer education and the policy environment among others. Stakeholders in the livestock sector have a role to play in enhancing the farmer's resilience capacity.

6.5 Conclusion

The study sought to examine the significance of livestock in building smallholder farmers' resilience against environmental and socio-economic shocks in ECRAS, ECRIMS and PROGRESS intervention areas in Zimbabwe. The results indicated

Table 6.8 Strategies to increase farmers' resilience to shocks

Shock	Strategy
Drought	Production of home-based animal feeds from locally available resources such as legumes, fodder conservation, conventional feeding, pen fattening, bush meal and urea treatment of poor quality stovers (UNDP, 2020a, b; Mudzengi et al., 2022)
Livestock theft	Investing in appropriate shelter and smart technologies to combat livestock theft; insurance against theft (Maluleke, 2022)
Low livestock prices	Improving livestock marketing; favourable policies; avoiding middlemen (Mondal et al., 2023)
Livestock death from diseases	Use of the services of community animal health workers, ethnoveterinary medicines, disease-resistant indigenous breeds (live bulls or artificial insemination) and insurance (Marpaung et al., 2022)

that livestock assets play a pivotal role in smallholder farmers' resilience to shocks. Generally, in semi-arid areas, livestock production is economically more viable than crop production. Livestock practices over a three-year period of surveys showed significant contribution to resilience in all consortia for all the four selected shocks (drought, livestock death due to diseases, livestock theft and low prices). In addition, owning donkeys or pigs significantly increased the resilience capacity index of households in the year 2019, whilst in 2020 and 2021, owning poultry, sheep and goats, cattle and rabbits significantly improved the resilience capacity of households. In fact, the type and numbers of livestock have a bearing on resilience capacity. Sale of livestock products has the capacity to build resilience, particularly during drought conditions, low livestock prices and livestock deaths from diseases, as revealed in ECRAS, ECRIMS and PROGRESS intervention areas. Routine health management practices such as vaccination, castration, deworming/dosing and dipping also improved livestock production and productivity. Therefore, adopting strategies that are particularly aimed at improving livestock production and productivity during shocks can further increase resilience of smallholder farmers. Other non-livestock assets also contributed towards farmers` resilience. Thus, non-livestock assets can be important fall-back plans to preserve livestock when faced with shocks, especially low market prices.

References

AL-Zoubi, R., Moussaoui, M., Abijaoude, R., Abo El Wafa, M., Ali, M., Chachaty, N., Galié, A., Hovhannisyan, A., Rajabovich, S., Wagha, A., Zaibet, L., Djalalov, S., & Tashmatov, A. (2009). Looking into the future for agriculture and AKST. In B. D. McIntyre, H. R. Herren, J. Wakhungu, & R. T. Watson (Eds.), *Agriculture at a Crossroads* (pp. 307–376). Island Press.

AU-IBAR. (2016). *VET-GOV; Livestock policy landscape in Africa: A review*.

Bahadur, A. V., Ibrahim, M., & Tanner, T. (2013). Characterising resilience: Unpacking the concept for tackling climate change and development. *Climate and Development, 5*(1), 55–65.

Bahta, Y. T. (2022). Nexus between coping strategies and households' agricultural drought resilience to food insecurity in South Africa. *Land, 11*(6), 893.

Bahta, Y. T., & Myeki, V. A. (2021). Adaptation, coping strategies and resilience of agricultural drought in South Africa: Implication for the sustainability of livestock sector. *Heliyon, 7*(11), e08280.

Berghof, T. V., Poppe, M., & Mulder, H. A. (2019). Opportunities to improve resilience in animal breeding programs. *Frontiers in Genetics, 9*, 692.

Cohn, A. S., Newton, P., Gil, J. D., Kuhl, L., Samberg, L., Ricciardi, V., Manly, J. R., & Northrop, S. (2017). Smallholder agriculture and climate change. *Annual Review of Environment and Resources*. https://doi.org/10.1146/annurev-environ-102016-060946

Dahwa, E., Mudzengi, C. P., Mubvuma, M., Maravanyika, T., Chapungu, L., & Chikodza, E. (2022). Optimizing productivity in semi-arid dryland agriculture for developing countries: Insights from Zimbabwe. In *Climate change adaptations in dryland agriculture in semi-arid areas* (pp. 233–249). Springer.

Field, C. B., Barros, V. R., Mach, K. J., Mastrandrea, M. D., van Aalst, R. A., Adger, W. N., Arent, D. J., Barnett, J., Betts, R. A., Bilir, T. E., & Birkmann, J. (2014). *Technical summary*. Cambridge University Press.

Food and Agriculture Organisation (FAO). (2016). *Characterization of the agricultural drought prone areas at a global scale*. http://www.fao.org/3/a-i5764e.pdf

Frischen, J., Meza, I., Rupp, D., Wietler, K., & Hagenlocher, M. (2020). Drought risk to agricultural systems in Zimbabwe: A spatial analysis of hazard, exposure, and vulnerability. *Sustainability, 12*(3), 752.

Gloede, O., Menkhoff, L., & Waibel, H. (2013). Shocks, individual risk attitude, and vulnerability to poverty among rural households in Thailand and Vietnam. *World Development, 71*, 54–78. https://doi.org/10.1016/j.worlddev.2013.11.005

Harvey, C. A., Saborio-Rodríguez, M., Martinez-Rodríguez, M. R., Viguera, B., Chain-Guadarrama, A., Vignola, R., & Alpizar, F. (2018). Climate change impacts and adaptation among smallholder farmers in Central America. *Agriculture & Food Security, 7*, 57. https://doi.org/10.1186/s40066-018-0209-x

Heltberg, R., & Lund, N. (2009). Shocks, coping, and outcomes for Pakistan's poor: Health risks predominate. *The Journal of Development Studies, 45*, 889–910. https://doi.org/10.1080/00220380902802214

Hennemuth, B., Bender, S., Bülow, K., Dreier, N., Keup-Thiel, E., Krüger, O., Mudersbach, C., Radermacher, C., & Schoetter, R. (2013). *Statistical methods for the analysis of simulated and observed climate data applied in projects and institutions dealing with climate change impact and adaptation*. CSC.

Herrero, S. G., Saldaña, M. Á. M., Rodriguez, J. G., & Ritzel, D. O. (2012). Influence of task demands on occupational stress: Gender differences. *Journal of Safety Research, 43*(5–6), 365–374.

Hurst, P., Termine, P., & Karl, M. (2005). *Agricultural workers and their contribution to sustainable agriculture and rural development*. FAO.

IPCC. (2014). *Climate change: Synthesis report*.

Jiri, O., Mtali-Chafadza, L., & Mafongoya, P. L. (2017). Influence of smallholder farmers' perceptions on adaptation strategies to climate change and policy implications in Zimbabwe. *Change and Adaptation in Socio-Ecological Systems, 3*(1), 47–55. https://doi.org/10.1515/cass-2017-0005

Kendall, M. G. (1975). *Rank correlation methods* (4th ed.). Charles Griffin.

Kom, Z., Nethengwe, N. S., Mpandeli, N. S., & Chikoore, H. (2020). Determinants of small-scale farmers' choice and adaptive strategies in response to climatic shocks in Vhembe District, South Africa. *GeoJournal, 87*, 677–700. https://doi.org/10.1007/s10708-020-10272-7

Lazzarini, I., Detmann, E., Sampaio, C. B., Paulino, M. F., de Campos Valadares, F., Augusto de Souza, M., & Oliveira, F. A. (2009). Intake and digestibility in cattle fed low-quality tropical forage and supplemented with nitrogenous compound. *Revista Brasileira de Zootecnia, 38*(10), 2021–2030.

Lipper, L., Thornton, P., Campbell, B. M., Baedeker, T., Braimoh, A., Bwalya, M., Caron, P., Cattaneo, A., Garrity, D., Henry, K., & Hottle, R. (2014). Climate-smart agriculture for food security. *Nature Climate Change*. https://doi.org/10.1038/nclimate2437

Lowder, S. K., Skoet, J., & Raney, T. (2016). The number, size, and distribution of farms, smallholder farms, and family farms worldwide. *World Development*. https://doi.org/10.1016/j.worlddev.2015.10.041

Mabuza, M. L., Ortmann, G. F., Wale, E., & Mutenje, M. J. (2016). The effect of major income sources on rural household food (in) security: Evidence from Swaziland and implications for policy. *Ecology of Food and Nutrition, 55*, 209–230.

Maluleke, W. (2022). Combating stock theft with deoxyribonucleic acid samples and innovative technology in South Africa. *International Journal of Social Science Research and Review, 5*(4), 190–213.

Mandarino, R. A., Barbosa, F. A., Lopes, L. B., Telles, V., Florence, E. D. A. S., & Bicalho, F. L. (2019). Evaluation of good agricultural practices and sustainability indicators in livestock systems under tropical conditions. *Agricultural Systems, 174*, 32–38.

Mann, H. B. (1945). Nonparametric test against trend. *Econometrical, 13*, 245–259.

Marpaung, L. N., Dahriansah, D., & Maulana, C. (2022). Expert system diagnose diseases in goats using the application of certainty factor at the Asahan Districk livestock service. *Jurnal Teknik Informatika (Jutif), 3*(1), 169–176.

Matlou, R., Bahta, Y. T., Owusu-Sekyere, E., & Jordaan, H. (2021). Impact of agricultural drought resilience on the welfare of smallholder livestock farming households in the northern cape province of South Africa. *Land, 10*(6), 562.

Mdungela, N. M., Bahta, Y. T., & Jordaan, A. J. (2017). Indicators for economic vulnerability to drought in South Africa. *Development in Practice, 27*(8), 1050–1063.

Mondal, M., Biswas, A., Mandal, S., Bhattacharya, S., & Paul, S. (2023). Developing micro level resilience index for Indian Sundarban adopting resilience indicators for measurement and analysis (RIMA) methodology. *Geosystems and Geoenvironment*, 100129. https://doi.org/10.1016/j.geogeo.2022.100129

Mudzengi, C. P., Dahwa, E., & Kapembeza, C. S. (2022). A strategic livestock feeding framework for semi-arid areas. In *Climate change adaptations in dryland agriculture in semi-arid areas* (pp. 291–299). Springer.

Murungweni, C., Andersson, J., van Wijk, M. T., Gwitira, I., & Giller, K. (2012). Zhombwe (*Neorautanenia brachypus* (harms) c.A.Sm.)—A recent discovery for mitigating effects of drought on livestock in semi-arid areas of southern Africa. *Ethnobotany Research and Applications, 10*, 199–212.

Mushore, T. D., Mhizha, T., Manjowe, M., Mashawi, L., Matandirotya, E., Mashonjowa, E., Mutasa, C., Gwenzi, J., & Mushambi, G. T. (2021). Climate change adaptation and mitigation strategies for small holder farmers: A case of Nyanga District in Zimbabwe. *Frontiers in Climate, 3*, 676495. https://doi.org/10.3389/fclim.2021.676495

Ngenoh, E., Kebede, S., Bett, H., & Bokelmann, W. (2018). Coping with shocks and determinants among indigenous vegetable smallholder farmers in Kenya. *Agricultural Sciences, 9*, 804–823. https://doi.org/10.4236/as.2018.97057

Ngongondo, C. S., Xu, C. Y., Tallaksen, L. M., Alemaw, B., & Chirwa, T. (2011). Regional frequency analysis of rainfall extremes in southern Malawi using the index rainfall and L-moments approaches. *Stochastic Environmental Research and Risk Assessment, 25*(7), 939–955.

Ngongoni, N. T., Mapiye, C., Mupeta, B., Mwale, M., & Chimonyo, M. (2007). Potential of farm – produced crop residues as protein sources for small -medium yielding dairy cows. *African Journal of Agriculture, 2*, 309–317.

Paenda, O., Musemwa, L., Ndhleve, S., & Sibanda, M. (2020). Determinants of farmers' marketing choices and preferences under communal cattle farming: Evidence from Mwenezi District in Zimbabwe. *Journal of Human Ecology, 72*(1–3), 13–23. https://doi.org/10.31901/24566608.2020/72.1-3.3263

Poshiwa, X., Groeneveld, R. A., Heitkönig, I. M. A., Prins, H. H. T., & van Ierland, E. C. (2013). Wildlife as insurance against rainfall fluctuations in a semi-arid savanna setting of southeastern Zimbabwe. *Tropical Conservation Science, 6*(1), 108–125.

Rojas-Downing, M. M., Nejadhashemi, A. P., Harrigan, T., & Woznicki, S. A. (2017). Climate change and livestock: Impacts, adaptation and mitigation. *Climate Risk Management, 16*, 145–163. https://doi.org/10.1016/j.crm.2017.02.001

Sekhonyana, T. E., & Fulpagare, Y. G. (2015). Effects of urea treated maize Stover silage on growth performance of crossbred heifers. *Journal of Agriculture and Veterinary Science, 8*(5), 58–62.

Serrato-Corona, J. S., Rivas, A. L., & Perez, E. C. (2010). Effects of undegradable intake protein on milk yield bun and dry matter digestibility in lactating goats fed ammoniated corn Stover. *Journal of Animal and Veterinary Advances, 9*, 1761–1764.

Thornton, P. K. (2010). Livestock production: Recent trends, future prospects. *Philosophical Transactions of the Royal Society B: Biological Sciences, 365*(1554), 2853–2867.

UNDP. (2020a). ZRBF resilience knowledge hub. In M. Langworthy, K. Fox, S. Martin, L. Woodson, T. Aziz, R. Al-haddad, E. Cuellar, C. Hein, & J. Stack (Eds.), *ZRBF outcome monitoring survey: Round two program learning report*.

UNDP. (2020b). ZRBF resilience knowledge hub. In C. Hoot, J. Stack, & T. Bidi (Eds.), *The role of Fodder in building resilience: A scoping study*.

UNDP. (2021). ZRBF resilience knowledge hub. In S. Martin, J. Stack, S. Nelson, & M. Langworthy (Eds.), *ZRBF outcome monitoring survey: Round three program learning report*.

Wilson, G. A. (2012). Community resilience, globalization, and transitional pathways of decision-making. *Geoforum, 43*(6), 1218–1231.

Wright, I. A., Tarawali, S., Blümmel, M., Gerard, B., Teufel, N., & Herrero, M. (2012). Integrating crops and livestock in subtropical agricultural systems. *Journal of the Science of Food and Agriculture, 92*(5), 1010–1015.

Zewdie, A. (2014). Impacts of climate change on food security: A literature review in sub Saharan Africa. *Journal of Earth Science and Climatic Change, 5*,8. https://doi.org/10.4172/2157-7617.1000225

Zimbabwe Population Census. (2012). *Provincial reports ZIMSTAT*. http://www.zimstat.co.zw/wp-content/uploads/publications/Population/population/Mat-North.pdf

Chapter 7
Improving Resilience Capacities of Vulnerable Farmers: Characteristics and Factors of People Who Have Moved Out of Poverty, Evidence from Rural Zimbabwe

Jeanette Manjengwa, Doreen Tsitsi Tirivanhu, and Lawrence Dhliwayo

7.1 Introduction

Globally, by the end of 2020, there were over 700 million people living in extreme poverty and 47%, almost half of the world's population, lives on less than US $6.85 a day (World Bank, 2022). The world's poor people are disproportionately vulnerable to loss of livelihood and assets, dislocation, hunger and famine in the face of climate variability and change (Cannon, et al., 2003). Most of the extreme poor live in Sub-Saharan Africa, with approximately two thirds living on less than US$2.15 a day and in addition, just over half of all households in Sub-Saharan Africa experienced multidimensional poverty (World Bank, 2022).

In Zimbabwe, income poverty is evaluated based on poverty thresholds established by the Zimbabwe National Statistics Agency (ZIMSTAT) at a spending level beneath which individuals or households experience absolute deprivation (ZIMSTAT, 2019). The total consumption poverty line (TCPL) represents the essential consumption expenditure needed to fulfil basic food and non-food necessities. Households considered 'poor' are those whose consumption of both food and non-food items falls below the total consumption poverty line but exceeds the food

J. Manjengwa (✉)
Department of Community and Social Development, University of Zimbabwe, Harare, Zimbabwe

D. T. Tirivanhu
Environment, Climate and Sustainable Development Institute, University of Zimbabwe, Harare, Zimbabwe

L. Dhliwayo
Department of Mathematics and Computation Sciences, University of Zimbabwe, Harare, Zimbabwe

L. Mhlanga et al. (eds.), *Building Resilience to Climate Change in Selected Districts in Zimbabwe*, https://doi.org/10.1007/978-3-031-80529-5_7

poverty line (FPL). Individuals falling below both the total consumption poverty line and the food poverty line are labelled as 'very poor,' facing extreme poverty and struggling to satisfy the minimum daily food requirements of 2100 calories.

Levels of poverty and extreme poverty in Zimbabwe remain unacceptably high. By 2017, according to ZIMSTAT's 'Zimbabwe Poverty Report 2017', which is based on the data derived from the Poverty, Income, Consumption and Expenditure Survey (PICES) 2017 survey results, national individual poverty levels were 70.5% and rural poverty levels were higher at 86.0% and had increased from 82.4% in 2001 (ZIMSTAT, 2019). In 2017, extreme poverty levels were 29.3% (ZIMSTAT, 2019). Extreme rural poverty levels increased from 30.4% in 2011 to 40.9% in 2017 (ZIMSTAT, 2019). World Bank estimated that by 2020, during the COVID-19 pandemic, extreme poverty had risen to 49.0% (World Bank, 2021).

The poverty situation is exacerbated by the multiple shocks and stresses faced by rural communities. Shocks include health shocks such as COVID-19, water-borne diseases, malaria, TB, HIV and AIDS; drought and the worsening climate crisis; the on-going macro-economic crisis and low industrial performance (Manjengwa 2021). The climate crisis, which has mainly manifested in the form of increased frequency of drought and various forms of extreme weather, has severe negative impacts on water security, health and food security (Government of Zimbabwe, 2014). Climate changes and variable weather patterns present hazards to individual households and communities. The impacts stemming from storms, droughts and gradual climate shifts are influenced by the social, political and economic vulnerabilities of individuals and communities on the grassroots level (Ribot, 2010). Climate change impacts Zimbabwe's agriculture-dependent economy, which heavily relies on rain-fed farming, livestock rearing and natural resources for livelihoods (Government of Zimbabwe, 2014). Thus, small-scale farmers, given their overreliance on climate-sensitive resources, are particularly vulnerable and their livelihoods precarious. The recurring shocks and stresses have long-term consequences for these households' livelihoods, particularly those in rural areas, which are left struggling to recover and primarily unable to expand economic opportunities or improve their livelihoods or well-being. People living in the rural areas have limited adaptive, adsorptive and transformational capacities, making them increasingly vulnerable to climate change (Ribot, 2010). Thus, building resilience should focus on addressing these socio-economic ills.

The notion of resilience has surfaced as a sustainable and cost-efficient framework to enhance the ability of susceptible communities to safeguard themselves and to attain and sustain improved well-being amidst recurring shocks and pressures (UNDP, 2015). Within this expanding global trend, Zimbabwe has witnessed a rise in financial and technical assistance from diverse development partners to bolster resilience initiatives and fortify the resilience capacities of vulnerable rural communities heavily reliant on rain-fed agriculture (UNDP, 2016). An example of such a programme is the Zimbabwe Resilience Building Fund Programme (ZRBF), which was executed in seven partnerships between 2015 and 2022 across 18 rural districts in Zimbabwe (Martin, et al., 2021). Established in response to various severe economic, environmental and social shocks impacting rural communities in Zimbabwe, the fund received co-funding from the UK's Foreign, Commonwealth and Development Office (FCDO), the European Union, the Embassy of Sweden and the Government of Denmark (Martin, et al., 2021). The management of the fund was overseen by the United Nations Development Programme (UNDP), with the Ministry of Lands, Agriculture, Fisheries, Water, and Rural Development serving as the primary implementation partner (Martin, et al., 2021). The programme's objective was to enhance resilience and bolster the capacity of vulnerable communities to shield themselves from shocks and stressors, striving to achieve and sustain improved well-being amidst challenges. Positioned as a pioneering resilience initiative in Zimbabwe, it is strategically positioned to extract insights for shaping future resilience programmes, providing evidence to inform relevant policies geared towards enhancing resilience in Zimbabwe, and contributing to the development of local capabilities to meet the country's national development objectives (UNDP, 2016).

The ZRBF programme focuses on building the three resilience capacities of absorptive, adaptive and transformative capacity with the long-term goal of improving food security and nutrition and well-being of beneficiaries (UNDP, 2015). Absorptive capacity minimises exposure to shocks and stresses, while adaptive capacity involves informed choice on alternatives to adapt to the situation. The transformative capacity comprises the governance structures, policies and regulations, infrastructure, community networks as well as formal and informal social protection mechanisms that collectively form the supportive framework for systemic transformation (Murima and Manyena, 2014).

While the Zimbabwe Resilience Building Fund (ZRBF) holds promise for enhancing the livelihoods of impoverished rural communities in developing nations like Zimbabwe, which are susceptible to shocks and pressures, there exists a dearth of local research to determine the impact of resilience capabilities on the socio-economic status of households, particularly their poverty levels. Consequently, this study delves into the influence of resilience capacity and the three resilience pillars on diverse socio-economic strata within beneficiary households. Leveraging the experiences of households benefiting from the ZRBF programme, the research aims to gain deeper insights into the role of resilience capacity development as a viable avenue for mitigating the vulnerability of economically disadvantaged rural farmers and understanding the attributes of beneficiary households making sustained strides out of poverty. More specifically, the study determined the poverty levels of the ZRBF recipient households, explored the relationship between social-economic status (extremely poor, poor and non-poor) and resilience capacities of rural households involved in the ZRBF Programme and determined the characteristics and factors of the beneficiary households that have both increased resilience capacity and improved socio-economic status.

7.2 Background

Zimbabwe is still largely a rural society with 61.4% of its people living in dispersed rural settlements, including communal lands, resettlement schemes and commercial farming areas, depending on agriculture and other land-based livelihoods (ZIMSTAT, 2022). Rural areas in Zimbabwe have generally experienced limited development, with poverty rates surpassing the national average. The origins of poverty in Zimbabwe can be traced back to the economic and conservation disparities entrenched in colonial policies. The inequitable distribution of land along racial lines led to overcrowding in the communal areas, where 51% of the population resides (ZIMSTAT, 2019). Rural livelihoods primarily revolve around agriculture and rely on natural resources such as climate, soil, water and biodiversity. Studies have shown that households situated in agro-ecological regions IV and V, characterised by lower agricultural potential, exhibit notably reduced consumption expenditures compared to those in regions with relatively higher potential in agro-ecological regions I, II and III (Poulton, et al., 2002; Bird, 2010; and Manjengwa, et al., 2012). These disparities underscore the significant challenge of limited alternative livelihood options for households in vulnerable areas, predisposing them to enduring poverty and extreme deprivation (Manjengwa, et al., 2012).

Insufficient productivity serves as a poverty cycle for most farmers in the rural regions of Zimbabwe, where agriculture is primarily subsistence-based, leading to elevated poverty rates. Agricultural production is below optimum, and productivity is still low with maize yields at less than one tonne per hectare. For example, in the 2017/18 season, the national maize yield was 0.99 tonnes per hectare, while for Communal Areas it was significantly lower at 0.54 tonnes per hectare (Government of Zimbabwe, 2018). Maize yields have declined since the 1990s when levels were over 1.2 tonnes per hectare (Binswanger-Mkhize and Moyo, 2012). This decrease not only mirrors the significant depletion of soil nutrients but also the remarkably low labour productivity and meagre returns from rain-fed agriculture in Zimbabwe (Binswanger-Mkhize and Moyo, 2012).

In addition, many vulnerable smallholder farmers fail to cultivate all their land due to lack of rain or unreliable rainfall patterns, lack of draught power (due to cattle deaths), limited ploughing implements, inadequate seed or fertilizer and labour constraints, especially for weeding (Manjengwa, et al., 2012; Chronic Poverty Advisory Network (CPAN), 2022; Bird et al., 2023). Despite achieving favourable yields, farmers encountered challenges in marketing, leading to diminished profitability within their enterprises (CPAN, 2022; Bird et al., 2023).

While poverty manifests in various forms, the most severe instances involve the lack of food and essential non-food items, termed as food and total consumption poverty, respectively (Manjengwa, et al., 2016). Income poverty directly affects other dimensions of poverty. The major sources of household income are from either employment, agricultural production and sale, remittances, cash transfers, rents or pensions. Remittances enable investment in human capital formation and for some the accumulation of assets and investment in agriculture and enterprise and a sustained escape from poverty. Remittances can stimulate local economies, as they boost local demand for goods and services (CPAN, 2022).

The impact of climate change, evident in the rising occurrence and intensity of extreme weather events like persistent droughts, hailstorms, floods and other natural calamities, persists in afflicting rural communities and intensifying poverty among those with limited resilience (UNDP, 2016; Manjengwa, 2021). The non-poor, but vulnerable families are driven into poverty, while the poor categories become very poor and destitute. It is only the wealthier non-poor households that often have non-farm sources of livelihoods and savings that can withstand these extreme weather events and have some degree of resilience capacity (CPAN, 2022; Bird et al., 2023). Therefore, it is important to investigate what attributes and characteristics make households better off, moving out of poverty and becoming more resilient to cope with shocks.

7.3 Overview of the Research Area and Methodology

7.3.1 Study Area

The research area encompassed the seven ZRBF Consortia operating across 18 districts in Zimbabwe. These districts were selected based on an assessment of various hazards impacting rural livelihoods in the country (UNDP, 2016). Nine hazards or stressors were identified in Zimbabwe, including drought, mid-season dry spells, floods, landmines, fluctuations in cereal and livestock prices, crop pests and diseases, animal diseases, HIV and AIDS and diarrheal diseases (UNDP, 2016). Examining the spatial distribution and overlap of these hazards offered insights into their potential cumulative impact on livelihoods in rural areas of Zimbabwe (UNDP, 2016). The mapping exercise aimed to support the UNDP's efforts in designing the Building Resilience in Zimbabwe initiative: Towards a Resilience Strategic Framework, and involved conducting analyses at the district and ward levels through consultations with government, non-governmental and private sector stakeholders.

Each Consortium comprised households in targeted Wards across multiple districts. The Consortia included: the Building Resilience through improving the Absorptive and Adaptive Capacity for Transformation (BRACT), operational in Nyanga, Mudzi and Mutoko Districts; the Enhancing Community Resilience and Sustainability (ECRAS) Consortium, active in Chiredzi and Mwenzi Districts; the Enhancing Community Resilience and Inclusive Market Systems (ECRIMS) Consortium, spanning Zvishavane and Mberengwa Districts; the Matabeleland Enhanced Livelihood Agriculture and Nutrition Adaptation (MELANA) in Nkayi, Bubi, Umzingwane and Umguza Districts; the Programme for Growth and Resilience (PROGRESS) in Beitbridge District; Action for Building Resilience in Zimbabwe (SIZIMELE) across Matobo, Lupane and Insiza Districts; and the Zambezi Valley Alliance for Building Community Resilience (ZVA) in Mbire, Binga and Kariba Districts.

7.3.2 Research Methods

7.3.2.1 Data Sources and Collection Methods

The research employed quantitative secondary data gathered during the ZRBF Outcome Monitoring Survey (OMS) spanning from 2019 to 2021. A mixed-methods approach was adopted to track ZRBF top-line metrics and assess the effects of interventions on outcomes, involving three rounds of a beneficiary-focused OMS each year from 2019 onwards (Martin, et al., 2021). The primary goals of the OMS were to execute a beneficiary-centred longitudinal survey to furnish outcome metric estimates for ZRBF's yearly reports and to evaluate the impact of engaging in project-backed initiatives and the subsequent enhancements in household and community resilience capacities concerning indicators amidst climate, economic and other adversities.

Information was gathered from over 3000 households concerning 11 ZRBF top-line indicators, encompassing outcome aspects like the utilisation or adoption of project-supported activities and impact factors such as household earnings, food security and nutrition. One of the OMS indicators pertains to outcome 4: Average monthly household income or a proxy for income.

The unit of measurement is average (in USD) per household per month. The OMS is disaggregated by consortium, household head sex and programme intensity. Per-capita monthly income and expenditures were regarded as economic well-being indicators for the regression analysis with resilience capacities and shock exposures.

The OMS therefore provide a wealth of data and information to investigate the role of resilience capacity building in reducing the vulnerability of resource poor rural farmers and ways in which they can achieve sustained escapes out of poverty.

The OMS is divided into two stratifications depending on the extent of programming. As outlined in the study (ZRBF, 2019), a two-phase cluster sampling approach was implemented, where wards were treated as clusters, with ten wards chosen from both the medium intensity and high intensity sampling frameworks for each consortium. Medium intensity involvement indicates that beneficiaries engaged in two or fewer ZRBF project activities, whereas high intensity signifies participation in three or more project activities. Table 7.1 presents the consortium, number of districts, number of wards, sampling frame and samples interviewed.

The study utilised data from all seven of the ZRBF Consortia specifically BRACT, ECRAS, ECRIMS, MELANA, PROGRESS, SIZEMELE and ZVA.

To compute various resilience capacity indices such as absorptive, adaptive and transformative, multivariate statistical analysis techniques including factor analysis and principal component analysis (PCA) were employed, as detailed in ZRBF (2019). The composite resilience capacity index was calculated by averaging the

Table 7.1 Number of beneficiaries by consortium (ZRBF, 2019)

				# Of beneficiaries								
		# Of wards (clusters)		In the sampling frame			Estimated sample size			Interviewed		
Consortium	# Of districts	Total	Selected	High	Med	Total	High	Med	Total	High	Med	Total
BRACT	2	28	20	2070	7803	9873	300	300	600	264	253	517
ECRAS	2	29	20	2723	6958	9681	300	300	600	259	241	500
ECRIMS	2	50	20	2587	21,077	23,664	300	300	600	244	256	500
MELANA	4	58	20	8690	12,566	21,256	300	300	600	209	240	449
PROGRESS	2	34	20	1919	13,077	15,796	300	300	600	249	234	483
SIZIMELE	3	57	20	12,627	15,595	28,222	300	300	600	249	257	506
ZVA	3	48	20	10,322	5,527	15,849	300	300	600	247	238	485
Total	IS	304	140	40,938	83,403	124,341	2,100	2,100	4,200	1,721	1,719	3,440

Source: ZRBF (2019)

standardised scores from PCA across 20 absorptive, adaptive and transformative resilience capacity elements. This calculation can be expressed as follows:

$$X_{sj} = \frac{X_j - X_{min}}{X_{max} - X_{min}} \times 100$$

where X_{sj} represents the standardised score value of component j, X_j stands for the actual score of the same component, and X_{min} and X_{max} denote the minimum and maximum values of component j, respectively.

7.3.3 *Analytical Framework*

Relationships between the resilience capacity indices such as absorptive, adaptive and transformative as presented in ZRBF (2019) were established and factor analysis was used to establish the characteristics or factors of households with improved well-being status. Further statistical data analyses, such as correlation and regression analysis, were performed to assess the cause and effect between different variables.

As discussed in the literature review, income poverty is measured against poverty lines set by ZIMSTAT at an expenditure level below which people or households are absolutely deprived (FPL and TCPL). These income and expenditure categorical variables set by ZIMSTAT were used to categorise data from ZRBF to be used for our statistical analysis. This was used in conjunction with resilience capacities obtained from ZRBF dataset.

For purposes of this study, we performed Chi-square test of relationship to determine the association between socio-economic status (extremely poor, poor and non-poor) and resilience capacities of rural households involved in the ZRBF programme. Furthermore, multinomial logistic regression was used to establish characteristics and factors that have improved both resilience capacity and socio-economic status of both beneficiary and non-beneficiary households.

Statistical analysis using the intervention analysis method was performed to determine the resilience pathways or resilience intervention of the sustained poverty escapers. These statistical methods, together with logistic regression, help us to determine whether we can attribute improved socio-well-being status to interventions introduced through the ZRBF programme.

7.3.4 *Data Analysis*

Statistical software, SPSS version 28 and STATA were used to analyse the data. Inferential and descriptive statistics were used to determine the change over time in household income as a proxy for socio-economic status. ZIMSTAT poverty datum lines were used to categorise household income into extremely poor, poor and non-poor. This is because the national statistics utilise ZIMSTAT definitions when reporting poverty. Statistical methods such as factor analysis, Chi-square test of relationship, logistics regression, multinomial regression analysis and intervention analysis were performed to analyse our data to achieve the aims and objectives of our research.

7.4 Results and Discussion

7.4.1 Poverty Lines

ZIMSTAT determines the total consumption poverty line (TCPL) (upper line) and the food poverty line (FPL) monthly, representing the expenses of a specific basket of goods for that period. Following ZIMSTAT's criteria, the FPLs and TCPLs were computed for the duration of the three years when the Zimbabwe Resilience Building Fund Programme (ZRBF) conducted its Outcome Monitoring Survey (OMS) (Table 7.2). Using the prices of the basket of goods and the various exchange rates pertaining to that period, Table 7.2 indicates that the TCPLs and FPLs fluctuated from year to year.

Table 7.2 shows that the FPL increased from USD 19.53 in 2019 to USD 47.78 in 2021, and the TCPL increased from USD 56.71 to USD 62.94 during the same period. During this period, there was accelerating inflation, use of multiple currencies and an economic crisis being experienced (Brien, et al., 2021). In February 2019, the Government implemented a new domestic currency, discontinuing the use of multiple currencies as legal tender. Despite these changes, they failed to halt the rapid increase in the official inflation rate, which surged to 838% in July 2020 before decreasing to 194% by April 2021 (Reserve Bank of Zimbabwe in CGTN, 2021). In early 2021, the Government temporarily reintroduced the United States dollar and South African Rand as legal tender and initiated a foreign exchange auction system. This measure helped stabilise the depreciation of the local currency (Brien, et al., 2021).

Table 7.2 ZIMSTAT's food and total consumption poverty lines for 2019, 2020 and 2021

Date		ZWL$	Exchange rate	US$
April 2019	FPL	63.72	3.2614	19.53
	TCPL	184.96	3.2614	56.71
March 2020	FPL	419.42	25.0000	16.78
	TCPL	1058.53	25.0000	42.34
March 2021	FPL	4033.27	84.4001	47.78
	TCPL	5312.19	84.4001	62.94

7.4.2 *Mean Poverty Lines for the ZRBF Consortia*

The mean food and total consumption poverty lines were calculated from the OMS data for each of the three rounds of data collection. Table 7.3 shows the FPLs for the three OMS rounds for the ZRBF.

The mean expenditure of food for the ZRBF Consortia for round 1 was USD 20.89 and is greater than the FPL value of USD 19.54 (Martin et al., 2021). This means that, on average, individuals in the sample were not extremely poor, that is they were therefore either poor or non-poor, on average.

Similarly, for round 2, the mean expenditure for food was USD 19.38 and is greater than ZIMSTAT's FPL of USD 16.78 (Martin et al., 2021). This implies that, on average, individuals were not extremely poor, but were either poor or non-poor.

However, by round 3, the mean expenditure for food was USD 31.02 and was less than ZIMSTAT's FPL of USD 47.79, that is, on average, individuals were extremely poor.

One-way analysis of variance was used to compare whether there is no significant difference in mean expenditure of food consumption. The probability value of 0.000 indicated that there was significant difference from round to round. We performed Post Hoc tests to identify which ones are different. All the probability values were less than 0.05, which indicated that the expenditure of food consumption was different from each other for all the three rounds.

Regarding the TCPL, monthly expenditure for both food and non-food expenditure for round 1 was US$ 46.22 and was lower than ZIMSTAT's TCPL value of US$ 56.71. This means that most people were in the poor or extremely poor category. Similarly, for round 2, mean monthly expenditure was US$ 40.96 and was lower than TCPL value of US$ 66.49, which means that most people are in the poor or extremely poor category. For round 3, expenditure was found to be US$ 64.84, which is higher than the TCPL value of US$ 62.94. This means that in round 3, people, on average, were in the non-poor category.

One-way analysis of variance was conducted to see whether there are differences in mean monthly expenditure from round to round. We found that the probability value was 0.000, which shows that that the mean expenditures are different from round to round. A Post hoc test was also performed to see which rounds were different. The probability values are all 0.000, meaning that mean differences of expenditure are all significant.

Table 7.3 Mean food consumption poverty (extreme poverty) for the ZRBF consortia in (USD)

OMS survey rounds	Mean	FPL	N	Std. deviation
1—April 2019	20.8944	19.54	4198	22.40759
2—March 2020	19.3827	16.78	4306	33.64353
3—March 2021	31.0245	47.79	4366	31.79429
Total	23.8251		12,870	30.20373

7.4.3 *Poverty Classification of the ZRBF Consortia*

Table 7.4 indicates the poverty classifications of extremely poor (that is below the food poverty line), the poor and non-poor for each of the three OMS rounds.

Table 7.4 shows that there was an increase in those classified as extremely poor from 29.5% (2564 households) in round 1 to 38.8% (3373 households) in round 3. For those classified as poor, there was a drastic decrease from 54.0% (964 households) in round 1 to 1.7% (31 households) in round 3. For the non-poor there was an increase from 28.7% (645 households) in round 1 to 40.6% (914 households) in round 3. This might be an indication that ZRBF intervention is effective in some cases. However, some of those previously classified as poor would have moved downwards to extremely poor category. In addition, overall, the number of households in the very poor category remains the largest at 3373.

This same pattern can be observed, without loss of generality, when we analyse by consortium. However, the pattern was variable across the consortia (Table 7.5).

Table 7.5 illustrates that there were differences in the performances of the seven ZRBF Consortia regarding poverty trajectories over the three rounds of the survey. There was an increase in the non-poor percentage over the three rounds for three Consortia: ZVA, ECRIMS and BRACT. There was little change in the non-poor percentage over the three rounds for MELANA, PROGRESS and SIZIMELE. One Consortium, ECRAS, had a decline in non-poor percentage over the three rounds.

This study looked at the link between participation in the ZRBF resilience programmes and activities and poverty categories, with the expectation that participation would improve overall well-being. Based on our analysis, in round 1 carried out in 2019, most people were in the poor category with an average expenditure between FPL and TCPL. However, this was only slightly above the extremely poor category of FPL expenditure. Those in round 2, that is March 2020, were also in the poor category with an average expenditure also between FPL and TCPL. In round 3

Table 7.4 Cross-tabulation of poverty classification for the three OMS survey rounds

			OMS survey rounds			
			Round 1	Round 2	Round 3	Total
Poverty classification	Extreme poor	Count	2564	2766	3373	8703
		% within poverty classification	29.5%	31.8%	38.8%	100.0%
	Poor	Count	964	790	31	1785
		% within poverty classification	54.0%	44.3%	1.7%	100.0%
	Non-poor	Count	645	692	914	2251
		% within poverty classification	28.7%	30.7%	40.6%	100.0%
Total		Count	4173	4248	4318	12,739
		% within poverty classification	32.8%	33.3%	33.9%	100.0%

Table 7.5 Cross-tabulation of poverty classification for each ZRBF consortium for the three OMS survey rounds

Consortium partner name				OMS survey rounds			Total
				Round 1	Round 2	Round 3	
1. BRACT	Poverty classification	Extreme poor	Count	264	308	321	893
			% within poverty classification	29.6%	34.5%	35.9%	100.0%
		Poor	Count	150	104	5	259
			% within poverty classification	57.9%	40.2%	1.9%	100.0%
		Non-poor	Count	54	63	151	268
			% within poverty classification	20.1%	23.5%	56.3%	100.0%
	Total		Count	468	475	477	1420
			% within poverty classification	33.0%	33.5%	33.6%	100.0%
2. ECRAS	Poverty classification	Extreme poor	Count	167	165	280	612
			% within poverty classification	27.3%	27.0%	45.8%	100.0%
		Poor	Count	81	85	0	166
			% within poverty classification	48.8%	51.2%	0.0%	100.0%
		Non-poor	Count	96	115	84	295
			% within poverty classification	32.5%	39.0%	28.5%	100.0%
	Total		Count	344	365	364	1073
			% within poverty classification	32.1%	34.0%	33.9%	100.0%
3. ECRIMS	Poverty classification	Extreme poor	Count	238	250	282	770
			% within poverty classification	30.9%	32.5%	36.6%	100.0%
		Poor	Count	128	110	1	239
			% within poverty classification	53.6%	46.0%	0.4%	100.0%
		Non-poor	Count	106	122	199	427
			% within poverty classification	24.8%	28.6%	46.6%	100.0%
	Total		Count	472	482	482	1436
			% within poverty classification	32.9%	33.6%	33.6%	100.0%

(continued)

Table 7.5 (continued)

Consortium partner name				OMS survey rounds Round 1	Round 2	Round 3	Total
4. MELANA	Poverty classification	Extreme poor	Count	470	514	601	1585
			% within poverty classification	29.7%	32.4%	37.9%	100.0%
		Poor	Count	153	132	7	292
			% within poverty classification	52.4%	45.2%	2.4%	100.0%
		Non-poor	Count	62	35	83	180
			% within poverty classification	34.4%	19.4%	46.1%	100.0%
	Total		Count	685	681	691	2057
			% within poverty classification	33.3%	33.1%	33.6%	100.0%
5. PROGRESS	Poverty classification	Extreme poor	Count	277	289	363	929
			% within poverty classification	29.8%	31.1%	39.1%	100.0%
		Poor	Count	98	72	3	173
			% within poverty classification	56.6%	41.6%	1.7%	100.0%
		Non-poor	Count	131	144	134	409
			% within poverty classification	32.0%	35.2%	32.8%	100.0%
	Total		Count	506	505	500	1511
			% within poverty classification	33.5%	33.4%	33.1%	100.0%
6. SIZIMELE	Poverty classification	Extreme poor	Count	513	537	813	1863
			% within poverty classification	27.5%	28.8%	43.6%	100.0%
		Poor	Count	263	198	11	472
			% within poverty classification	55.7%	41.9%	2.3%	100.0%
		Non-poor	Count	139	168	157	464
			% within poverty classification	30.0%	36.2%	33.8%	100.0%
	Total		Count	915	903	981	2799
			% within poverty classification	32.7%	32.3%	35.0%	100.0%

(continued)

Table 7.5 (continued)

Consortium partner name				OMS survey rounds Round 1	Round 2	Round 3	Total
7. ZVA	Poverty classification	Extreme poor	Count	634	705	715	2054
			% within poverty classification	30.9%	34.3%	34.8%	100.0%
		Poor	Count	91	90	3	184
			% within poverty classification	49.5%	48.9%	1.6%	100.0%
		Non-poor	Count	57	44	106	207
			% within poverty classification	27.5%	21.3%	51.2%	100.0%
	Total		Count	782	839	824	2445
			% within poverty classification	32.0%	34.3%	33.7%	100.0%

(March 2021) of the survey, most people were in the extremely poor category with an average expenditure below FPL and TCPL. Thus, people have moved from being poor in 2019 and 2020 to extremely poor in 2021. This can be attributed, without loss of generality, to COVID-19, because of lockdown.

This finding from the analysis of the survey data reflects Martin, et al., (2021), who indicted people's perception of having lost income over the course of the OMS rounds, A key factor that affected well-being during this period was the COVID-19 pandemic. COVID-19-related restrictions on travel and market and business closures, negatively impacted livelihoods, increased unemployment and reduced incomes (CPAN, 2022). During this timeframe, there was notable macroeconomic instability marked by cash scarcities, elevated inflation rates, credit limitations, various currencies and fluctuating exchange rates, resulting in unpredictable expenditure and income projections (CPAN, 2022).

The COVID-19 pandemic overlaid and compounded pre-existing crisis conditions in Zimbabwe, and multiple and sequenced shocks and negative trends have eroded resilience, pushing many into adverse and erosive forms of coping (CPAN, 2022). The drought of 2019/2020 worsened the situation.

Concerning the repercussions of COVID-19 pandemic restrictions on vulnerable smallholder farmers, some found themselves pushed into distressing coping mechanisms, faced downward mobility, and, to this point, have been unable to regain their footing (CPAN, 2022). Unlike farmers who did not need to sell off their livestock, these individuals were unable to replenish their stocks. The limited market access stemming from lockdown measures led to produce spoilage and income loss.

7.4.4 Poverty Classification and Resilience Capacities

The study investigated the relationship between social-economic status (extremely poor, poor and non-poor) and resilience capacities of rural households involved in the ZRBF Programme. Calculations were done to cross tabulate the various resilience capacities (absorptive, adaptive and transformative and overall resilience capacity) and poverty categories.

7.4.5 Poverty and Absorptive Capacity

Table 7.6 shows the absorptive capacity according to poverty classification for the three rounds of OMS.

Table 7.6 Cross-tabulation of poverty classification for absorptive capacity of the three OMS survey rounds

OMS survey rounds				Absorptive capacity		Total
				No	Yes	
1	Poverty classification	Extreme poor	Count	2564		2564
			% within poverty classification	100.0%		100.0%
		Poor	Count	964		964
			% within poverty classification	100.0%		100.0%
		Non-poor	Count	645		645
			% within poverty classification	100.0%		100.0%
2	Poverty classification	Extreme poor	Count	2766		2766
			% within poverty classification	100.0%		100.0%
		Poor	Count	790		790
			% within poverty classification	100.0%		100.0%
		Non-poor	Count	692		692
			% within poverty classification	100.0%		100.0%
3	Poverty classification	Extreme poor	Count	1394	1980	3374
			% within poverty classification	41.3%	58.7%	100.0%
		Poor	Count	13	17	30
			% within poverty classification	43.3%	56.7%	100.0%
		Non-poor	Count	146	769	915
			% within poverty classification	16.0%	84.0%	100.0%

There were no absorptive capacities recorded in round 1 and round 2. In round 3, the extreme poor who said yes to absorptive capacity were 58.7%. The poor who said yes to absorptive capacity were 56.7%. The non-poor who said yes to absorptive capacity were 84.0%. This shows an increase in absorptive capacity in round 3 compared to round 1 and round 2. In round 3, the non-poor have the highest score (84%) for absorptive capacity. The extreme poor and poor had lower score at 58.7% and 56.7%, respectively. However, the share of the poor beneficiaries with absorptive capacity is larger than those without.

7.4.5.1 Poverty Classification and Adaptive Capacity

Table 7.7 shows the adaptive capacity according to poverty classification, for each of the survey rounds.

Table 7.7 Cross-tabulation of poverty classification and adaptive capacity for the OMS survey rounds

OMS survey rounds				Adaptive capacity No	Adaptive capacity Yes	Total
1	Poverty classification	Extreme poor	Count	2564		2564
			% within poverty classification	100.0%		100.0%
		Poor	Count	964		964
			% within poverty classification	100.0%		100.0%
		Non-poor	Count	645		645
			% within poverty classification	100.0%		100.0%
2	Poverty classification	Extreme poor	Count	1568	1198	2766
			% within poverty classification	56.7%	43.3%	100.0%
		Poor	Count	292	498	790
			% within poverty classification	37.0%	63.0%	100.0%
		Non-poor	Count	137	555	692
			% within poverty classification	19.8%	80.2%	100.0%
3	Poverty classification	Extreme poor	Count	1862	1511	3373
			% within poverty classification	55.2%	44.8%	100.0%
		Poor	Count	16	15	31
			% within poverty classification	51.6%	48.4%	100.0%
		Non-poor	Count	218	696	914
			% within poverty classification	23.9%	76.1%	100.0%

In round 1, there were no adaptive capacities. In round 2, the extreme poor who said yes to adaptive capacity were 43.3%. The poor who said yes to adaptive capacity were 63.0%. The non-poor who said yes to adaptive capacity were 80.2%. This shows a general increase in adaptive capacity in round 2 compared to round 1.

In round 3, the extreme poor who said yes to adaptive capacity were 44.8%. The poor who said yes to adaptive capacity were 48.4%. The non-poor who said yes to adaptive capacity were 76.1%. This shows a general increase in adaptive capacity in round 3. However, Table 7.7 shows that between round 2 and round 3, there was a decline in absorptive capacity among the poor and non-poor. For the non-poor, this decline was from 80.2% in round 2 to 76.1% in round 3.

7.4.6 Poverty and Transformative Capacity

Table 7.8 shows the transformative capacity according to poverty classification.

Table 7.8 shows that in round 1 there were no transformative capacities. In round 2, the extreme poor who said yes to transformative capacity were 52.8%. The poor who said yes to transformative capacity were 63.7%. The non-poor who said yes to transformative capacity were 71.2%. This shows a general increase in transformative capacity in round 2 compared to round 1.

In round 3, the extreme poor who said yes to transformative capacity were 53.4%. The poor who said yes transformative capacity were 63.3%. The non-poor who said yes to transformative capacity were 69.4%. This shows that there was not much of an increase in transformative capacity in round 3. In fact, for the non-poor, transformative capacity decreased from 71.2% to 69.4%.

7.4.7 Poverty and Overall Resilience Capacity

Table 7.9 shows the resilience capacity according to poverty classification for the three OMS rounds.

In round 1, there were no resilience capacities. In round 2, the extreme poor who said yes to resilience capacity were 46.2%. The poor who said yes to resilience capacity were 64.7%, and the non-poor who said yes to resilience capacity were 80.3%. This shows a general increase in resilience capacity in round 2 compared to round 1.

In round 3, the extreme poor who said yes to resilience capacity were 49.6%. The poor who said yes to resilience capacity were 45.2%, and the non-poor who said yes to resilience capacity were 77.1%. This shows an increase in resilience capacity for the extreme poor and a decrease for the poor and non-poor in round 3.

Table 7.8 Cross tabulation of poverty classification and transformative capacity for the three OMS survey rounds

OMS survey rounds				Transformative capacity		
				No	Yes	Total
1	Poverty classification	Extreme poor	Count	2564		2564
			% within poverty classification	100.0%		100.0%
		Poor	Count	964		964
			% within poverty classification	100.0%		100.0%
		Non-poor	Count	645		645
			% within poverty classification	100.0%		100.0%
2	Poverty classification	Extreme poor	Count	1307	1460	2767
			% within poverty classification	47.2%	52.8%	100.0%
		Poor	Count	287	503	790
			% within poverty classification	36.3%	63.7%	100.0%
		Non-poor	Count	199	492	691
			% within poverty classification	28.8%	71.2%	100.0%
3	Poverty classification	Extreme poor	Count	1573	1800	3373
			% within poverty classification	46.6%	53.4%	100.0%
		Poor	Count	11	19	30
			% within poverty classification	36.7%	63.3%	100.0%
		Non-poor	Count	280	634	914
			% within poverty classification	30.6%	69.4%	100.0%

7.4.8 Drivers of Poverty

Binary logistic regression was employed to investigate the potential associations between poverty classification, round, negative coping, consortium and the gender of the household head with the probability of possessing resilience capacity. The model demonstrated statistical significance (p-value = 0.000), indicating its ability to differentiate between individuals with and without resilience capacity. It accounted for approximately 40.3% (Cox and Snell R square) to 54.9% (Nagelkerke R-square) of the variability in the dependent variable and accurately classified 78% of cases.

As shown in Table 7.10, poverty classification, negative coping, consortium and sex of head of household significantly contributed to the model. Only round 1 did

Table 7.9 Cross tabulation of poverty classification and resilience capacity by the three OMS survey rounds

OMS Survey rounds				Resilience capacity		
				No	Yes	Total
1	Poverty classification	Extreme poor	Count	2564		2564
			% within poverty classification	100.0%		100.0%
		Poor	Count	964		964
			% within poverty classification	100.0%		100.0%
		Non-poor	Count	645		645
			% within poverty classification	100.0%		100.0%
2	Poverty classification	Extreme poor	Count	1489	1278	2767
			% within poverty classification	53.8%	46.2%	100.0%
		Poor	Count	279	512	791
			% within poverty classification	35.3%	64.7%	100.0%
		Non-poor	Count	136	556	692
			% within poverty classification	19.7%	80.3%	100.0%
3	Poverty classification	Extreme poor	Count	1700	1674	3374
			% within poverty classification	50.4%	49.6%	100.0%
		Poor	Count	17	14	31
			% within poverty classification	54.8%	45.2%	100.0%
		Non-poor	Count	209	705	914
			% within poverty classification	22.9%	77.1%	100.0%

not significantly contribute to the model. From Table 7.10 we can deduce the following:

- The odds ratio of having a resilience capacity are of those classified as extremely poor are 0.318 less than those which are classified as non-poor with a with a confidence interval of 0.277 to 0.365.
- The odds ratio of having a resilience capacity are of those classified as poor are 0.555 less than those which are classified as non-poor with a with a confidence interval of 0.453 to 0.681.
- The odds ratio of having a resilience capacity in round 3 are not significantly different from those in round 1 and round 2 since the p-value are all greater than 0.05 significance level.

The odds ratio of having a resilience capacity are of those in other consortiums are significantly different from those in consortium—BRACT as presented in Table 7.10.

Table 7.10 Variables in the equation

		B	S.E.	Wald	Df	Sig.	Exp(B)	95% C.I.for EXP(B)	
								Lower	Upper
Step 1[a]	Poverty_Class (non-poor)			286.567	2	0.000			
	Poverty_Class (extremely poor)	−1.146	0.070	266.575	1	0.000	0.318	0.277	0.365
	Poverty_Class (poor)	−0.588	0.104	31.923	1	0.000	0.555	0.453	0.681
	Round 3			0.846	2	0.655			
	Round 1	−21.963	588.586	0.001	1	0.970	0.000	0.000	.
	Round 2	−0.046	0.050	0.845	1	0.358	0.955	0.865	1.054
	Negative coping: Yes	0.281	0.072	15.235	1	0.000	1.325	1.150	1.525
	Consortium BRACT			745.836	6	0.000			
	ECRAS	1.104	0.138	63.667	1	0.000	3.015	2.299	3.954
	ECRIMS	0.635	0.114	30.960	1	0.000	1.887	1.509	2.359
	MELANA	−1.080	0.093	136.011	1	0.000	0.340	0.283	0.407
	PROGRESS	−0.921	0.099	87.017	1	0.000	0.398	0.328	0.483
	SIZIMELE	−0.973	0.088	122.936	1	0.000	0.378	0.318	0.449
	ZVA	−1.203	0.090	180.185	1	0.000	0.300	0.252	0.358
	hhh_sex female	−0.448	0.051	76.722	1	0.000	0.639	0.578	0.706
	Constant	1.675	0.115	210.821	1	0.000	5.338		

[a]Variable(s) entered in step 1: Poverty_Class, round, negative coping, consortium, hhh_sex

The odds ratio of having a resilience capacity are of those households headed by females is 0.639 less than those which are headed by males with a confidence interval of 0.578 to 0.706.

The extremely poor are less likely to have a high resilience capacity compared to non-poor since the odds ratio is 0.318 (less than 1). Similarly, the poor are less likely to have a high resilience capacity compared to non-poor since the odds ratio is 0.555 (less than 1). We cannot comment on the direction of the odds ratio of round 1 and round 2 compared to round 3 in terms of the resilience capacity since the odds p-values are less 0.05. Negative coping is more likely to have a high resilience capacity compared to those without negative coping since the odd ratio is 1.325 (more than 1).

ECRAS and ECRIMS are more likely to have a high resilience capacity compared to BRACT since the odd ratios are more than 1, that is 3.015 and 1.887, respectively. However, MELANA, PROGRESS, SIZIMELE and ZVA are less likely to have a high resilience capacity compared to BRACT since the odd ratios are less than 1, that is 0.340, 0.398, 0.378 and 0.300, respectively.

Female headed households are more likely to have a less resilience capacity compared to male headed households since the odds ratio is 0.639 (less than 1).

In summary, the logistic regression analysis indicates that individuals not classified as poor exhibit greater levels of absorptive, adaptive, transformative and overall resilience capacities compared to both poor and extremely poor individuals. The data collection rounds (round 1, round 2 and round 3) did not have a significant influence on these capacities. Negative coping mechanisms were associated with higher levels of absorptive, adaptive, transformative and resilience capacities when contrasted with non-negative coping strategies. ECRAS and ECRIMS are more inclined to possess high levels of absorptive, adaptive, transformative and resilience capacities compared to BRACT. Conversely, MELANA, PROGRESS, SIZIMELE and ZVA are less likely to exhibit high levels of absorptive, adaptive, transformative and resilience capacities in comparison to BRACT. Additionally, it is noteworthy that households headed by females are more inclined to have lower levels of absorptive, adaptive, transformative and resilience capacities compared to households headed by males.

7.5 Conclusion

This study examined the links between income poverty levels and resilience capacity to understand the role of resilience capacity building as a possible pathway for reducing the vulnerability of resource poor rural farmers. In particular, the study aimed to identify the characteristics of households who are moving out of poverty.

Zimbabwe is experiencing an ongoing economic crisis, high inflation and high levels of poverty (Sharma, et al., 2022). The COVID-19 pandemic increased poverty. We saw in Table 2 that ZIMSTAT's calculation of the food and total consumption poverty lines, using the prices of the basket of goods, fluctuated from year to year, due to inflation and multiple currency use. Our analysis found that in rounds 1 and 2, carried out in 2019 and 2020, most people were in the poor category. By March 2021, our analysis found that in round 3 most people were in the extremely poor category. Thus, people had moved from being poor in 2019 and 2020 to extremely poor in 2021.

Our analysis of the OMS survey results regarding resilience capacities found that in round 1 there were no resilience capacities. There was a general increase in resilience capacity in round 2 compared to round 1, whereas in round 3, the resilience capacity generally increased compared to round 2 and round 1. It is noted that the ZRBF programme was implemented between 2015 and 2022 by 7 consortia, across 18 rural districts in Zimbabwe. The ZRBF programme targeted interventions that sought to boost resilience amongst households and communities.

Analysis of the different resilience capacities, namely absorptive, adaptive and transformative, found that the picture regarding resilience and poverty levels was not so clear cut, and somewhat contradictory. Between round 2 and round 3, there was a decline in absorptive capacity among the poor and non-poor. For the non-poor, this decline was from 80.2% in round 2 to 76.1% in round 3. Results show that there was not much of an increase in transformative capacity in round 3. In fact, for the non-poor, transformative capacity decreased from 71.2% to 69.4%.

Binary logistic regression was used to examine whether the variables poverty classification, round, negative coping, consortium and sex of head of household were associated with the likelihood of having resilience capacity. The results show that only round 1 did not significantly contribute to the model, the rest of the variables contributed significantly. The extremely poor are less likely to have a high resilience capacity compared to non-poor since the odds ratio is 0.318 (less than 1). Similarly, the poor are less likely to have a high resilience capacity compared to non-poor since the odds ratio is 0.555 (less than 1).

Finally, we can conclude without loss of generality from logistic regression that non-poor people have more absorptive capacity, adaptive capacity, transformative capacity and resilience capacity than both poor and extremely poor people. Negative coping had a high absorptive capacity, adaptive capacity, transformative capacity and resilience capacity compared to non-negative coping. It is also important to note that female headed households are more likely to have a less absorptive capacity, adaptive capacity, transformative capacity and resilience capacity compared to male headed households.

7.6 Recommendations

The differing results amongst the seven Consortia highlight that there is a need for more investigation and research to determine the root causes of these differences, so as to tailor resilience building interventions to suit specific conditions and situations.

Therefore, there is a need to implement resilience programmes that target interventions that can have a lasting positive impact on the poor and extremely poor.

References

Binswanger-Mkhize, H., & Moyo, S. (2012). *Zimbabwe from economic rebound to sustained growth: Note II: Recovery and growth of Zimbabwe agriculture*. World Bank.

Bird, K. (2010). The intergenerational transmission of poverty: An overview. Chapter 4. In CPRC (Ed.), *Addressing chronic poverty: Concepts, cross-cutting issues and country contexts*. Routledge.

Bird, K., Diwakar, D., Manjengwa, J., Pindiriri, C., Chamunogwa, A., Ndlovu, N., Tirivanhu, T., Chevo, T., Mabeza, C., Chandaengerwa, E., Eichsteller, M., & Guigui, D. (2023). Understanding poverty dynamics in Zimbabwe in the context of Covid-19. In *National report*. CPAN, University of Zimbabwe and UK Aid.

Brien, S., Dugar, P., Herring, D., King, E., Montes, C., & Morrison, A. (2021). *Pathways to prosperity: Zimbabwe case study, overview*. Legatum Institute.

Cannon, T., Twigg, J., & Rowell, J. (2003). Social vulnerability, sustainable livelihoods and disasters. In *Report to the department for international development*. Conflict and Humanitarian Assistance Department and Sustainable Livelihoods Support Office.

CGTN. (2021). *Zimbabwe's central bank expects inflation to fall to 55 pct by July*. CGTN, 5 May 2021. (Accessed online 10/05/21).

Chronic Poverty Advisory Network (CPAN). (2022, May). Zimbabwe Covid-19 poverty monitor. In *CPAN and the environment, climate and sustainable development institute*. University of Zimbabwe. Covid Collective, IDS.

Government of Zimbabwe. (2014). *National climate change response strategy and action plans*. Ministry of Environment, Water and Climate.

Government of Zimbabwe. (2018, May). *Second Round Crop and Livestock Assessment Report, 7 9*. Ministry of Agriculture, Lands, Water and Rural Resettlement. .

Manjengwa, J. (2021). *Understanding poverty dynamics in Zimbabwe in the context of Covid-19: Literature review*. London:Chronic Poverty Advisory Network (CPAN).

Manjengwa, J., Feresu, S., & Chimhowu, A. (Eds.). (2012). *Understanding poverty, promoting wellbeing and sustainable development: A sample survey of 16 districts of Zimbabwe*. Institute of Environmental Studies, University of Zimbabwe.

Manjengwa, J., Tirivanhu, D., & Matema, C. (2016). Understanding urban poverty in two high density suburbs of Harare, Zimbabwe. *Development Southern Africa, 33*(1), 23–38.

Martin, S., Stack, J., Nelson, S. and M. Langworthy (2021). *Zimbabwe resilience building fund outcome monitoring survey round 3 2021 program learning report.*

Murima, B., & Manyena, B. (2014, September). *'Resilience framework for urban children and communities', presentation at the international conference 'towards resilient urban communities', rainbow towers*. University of Zimbabwe and UNICEF, Harare Programme Office.

Poulton, C., Davies, R., Matshe, I., & Urey, I. (2002). *A review of Zimbabwe's agricultural economic policies: 1980–2000*. Institutions and Economic Policies for Pro-Poor Agricultural Growth Project, DFID.

Ribot, J. C. (2010). Vulnerability does not just fall from the sky: Toward multi-scale pro-poor climate policy. In R. Mearns & A. Norton (Eds.), *Social dimensions of climate change: Equity and vulnerability in a warming world*. The World Bank.

Sharma, D., Alwang, J. R., Chingozha, T., Hoy, C., Kurasha, A., Nyash, F. M., & Paez Rodas, A. (2022). *Reversing the tide : Reducing poverty and boosting resilience in Zimbabwe – overview (English)*. World Bank Group.

UNDP. (2015, March). *Building resilience in Zimbabwe—Towards a resilience strategic framework*. Harare.

UNDP. (2016). *Mapping of selected hazards affecting rural livelihoods in Zimbabwe: A district and Ward analysis*. UNDP.

World Bank. (2021, June). Zimbabwe economic update. *Overcoming economic challenges, natural disasters, and the pandemic: Social and economic impacts*. Issue 3. The World Bank Group.

World Bank. (2022). *Poverty and shared prosperity 2022: Correcting course*. World Bank.

ZIMSTAT. (2019). *Zimbabwe poverty report 2017*. ZIMSTAT.

ZIMSTAT. (2022). *2022 population and housing census*. Preliminary report on population figures, ZIMSTAT.

ZRBF (Zimbabwe Resilience Building Fund). (2019). ZRBF resilience knowledge hub. In M. Langworthy, T. Aziz, & J. Stack (Eds.), *ZRBF outcome monitoring survey: Round one program learning report*. UNDP.

Chapter 8
Building Household Resilience to the Effects of Climate Change

Byron Zamasiya and Kefasi Nyikahadzoi

8.1 Introduction

Extreme events such as droughts adversely impact the livelihoods of smallholder farmers who rely on rain-fed agriculture for food and income security. The situation is further worsened by social and economic shocks and stresses, which strain the fallback mechanisms for the communities living on the climate frontier. This book has explored the varied impacts of extreme events on smallholder farmers. It examined their coping mechanisms and the consequences of these strategies on household food and income security. Drawing from the case studies, this chapter synthesises key findings and proposes policy interventions to bolster smallholder farmers' resilience against climate change and other extreme events. This chapter summarises the main findings from the case studies and proffers recommendations for enhancing the resilience of smallholder farmers to social, economic and environmental shocks and stresses.

8.2 Summary of Findings

The chapter by Nyika et al. shows that using quality seeds, small grains, crop rotation, improved varieties, mulching and integrated pest management are the most popular CSA practices among smallholder farmers in Zimbabwe. Results of the study show that sex, age, marital status, education, household size, frequency of climate shocks and the provision of agricultural information enhance CSA

B. Zamasiya (✉) · K. Nyikahadzoi
Environment Climate and Sustainability Institute, University of Zimbabwe, Harare, Zimbabwe
e-mail: bzamasiya@sociol.uz.ac.zw

L. Mhlanga et al. (eds.), *Building Resilience to Climate Change in Selected Districts in Zimbabwe*, https://doi.org/10.1007/978-3-031-80529-5_8

adoption. Their findings show that adopters of CSA practices have higher monthly incomes and are more food secure than non-adopters. The chapter further shows that adopting CSA practices has some positive returns on food security. Non-adopters of CSA practices are more food insecure than adopters. The chapter demonstrated that CSA practices are critical in building resilience in a changing climate as they lessen the impacts of events. They also help poor farmers to lessen the impacts of food price instability caused by other natural disasters such as the Covid-19 pandemic.

Zamasiya, Nyikahadzoi, and Mataruse found that several factors helped households cope with challenges (measured by the Coping Strategy Index). These included improved goat breeds; stronger social networks, access to financial services (though the impact was less certain), ownership of productive assets (like tools or equipment), higher levels of education and skills, access to information, strong connections to markets and other resources and ownership of livestock. Interestingly, the effectiveness of these factors depended on the specific type of shock a household faced.

The chapter by Bhatasara and Karuma demonstrates that communities experienced shocks such as droughts, excessive rains, contamination of freshwater sources and pasture and forage shortages. The study shows that the ZRBF programme established resilient infrastructures such as irrigation canals, mechanised boreholes, dams, fishponds and solar-powered boreholes. The chapter demonstrates that the programme provided water infrastructure close to households, increasing women's time to venture into other income-generating activities. Despite the provision of water infrastructure, there remained challenges to resilience, such as water diversion by households engaged in brickmaking and gold panning, environmental damage and costly repairs and maintenance.

Mwakiwa et al. studied how vulnerable smallholder communities cope with unexpected challenges. Their research explored the types of shocks these communities face, how they plan and respond and what factors influence their resilience. The study identified several key elements for building resilience, including the age of the household head, family size and income diversification and availability of services and information. The chapter demonstrates that assets and how households respond to shocks are essential to responding to hazards. The chapter further shows that social capital indices experienced substantial gains for the three years. Property titling and security of tenure, particularly ownership of homes, were shown to increase the likelihood of willingness to invest in longer-term adaptive mechanisms.

Mudzengi et al. examined the significance of livestock in building smallholder farmers' resilience against environmental and socio-economic shocks in ECRAS, ECRIMS and PROGRESS intervention areas in Zimbabwe. The chapter demonstrated that livestock assets play a pivotal role in smallholder farmers' resilience to shocks such as droughts. Livestock such as cattle play an important role in building farmers' resilience but are susceptible to shocks such as livestock theft, low prices and diseases. Other livestock assets such as chicken, sheep and goats played an important role in helping households withstand impact from droughts.

The chapter by Manjengwa, Tirivanhu and Dhliwayo examined the relationship between socioeconomic status and resilience capacities of rural households involved in the ZRBF programme, the characteristics and factors of the beneficiary households that have both increased resilience capacity and improved socio-economic status and the resilience pathways/ resilience interventions of the sustained poverty escapers. Their chapter demonstrated that, most households were in the extremely poor category, with people moving from being poor in 2019 and 2020 to extremely poor in 2021. Research results show that female-headed households were more likely to be less resilient. The extremely poor and poor households are less likely to have a high resilience capacity than the non-poor.

8.3 Discussion

8.3.1 Promote Gendered Design of Adaptation Technologies

Gender plays a critical factor in determining whether an adaptation technology will be adopted and if it will realise the desired impact in building the resilience of smallholder farmers. In Zimbabwe, women are the de jure household heads and the primary food producers. Nyika et al. have highlighted the importance of designing gender-sensitive CSA strategies to address the adaptation needs of smallholder farmers, most of whom are women. This can be done by conducting a gendered contextual analysis of the needs, power dynamics, socioeconomic realities, priorities and cultural realities of men and women before the commencement of the programme. The gender analysis will help recognise and adequately address the needs of men and women in the design and application of CSA to ensure that both men and women can equally benefit.

8.3.2 Enhance Livestock Asset Accumulation and Breeds

Livestock plays a vital role in building the resilience of smallholder farmers to environmental shocks such as droughts. Mudzengi et al. demonstrated that households that own cattle, pigs, sheep, poultry and goats were better able to withstand drought than their counterparts. In most dryland zones in Zimbabwe where the ZRBF programme was implemented, rainfall is highly variable, making it difficult to grow crops under rainfed conditions. Households in such areas rely on different forms of livestock for survival during environmental stresses such as droughts. The reliance on livestock is further supported by the fact that most rural areas in Zimbabwe are disconnected from the mainstream financial system. This leaves them to use livestock as a store of value, which they can draw down on during lean times. Zamasiya, Nyikahadzoi and Mataruse indicated that goats are helping smallholder farmers manage their food insecurity during environmental shocks. Goats thrive in regions with poor agroecological potential, such as Mbire, Kariba and Binga districts. The ZRBF baseline emphasised that these districts already provide distant markets for goats. Despite providing a safety net during lean times, livestock suffer from shocks such as theft and diseases (Mudzengi et al.). Price volatility also adversely affects farmers with large livestock such as cattle. Smallholder farmers participated in breed improvement under the ZRBF project to fetch better prices. This helped them to fetch higher incomes from improved breeds. With improved incomes from livestock sales, smallholder farmers can withstand the impacts of climate change. Therefore, promoting livestock asset accumulation and breed improvement is one of the critical factors that help improve smallholder farmers' resilience to shocks.

Livestock asset accumulation can be enhanced by targeting smallholder farmers with local cattle breeds and goat breeds. Artificial insemination can be promoted to cattle farmers to improve their breeds. This can be complemented with cattle fattening, where smallholder farmers can bring their cattle for fattening before selling. Cattle farmers with improved cattle breeds or fattened livestock could be linked to reliable buyers through contracts. Goat breeds can be improved by organising smallholder farmers into groups and providing them with a buck. The goat farmers can either bring their she-goats to one lead farmer for mating or rotate the buck among the members. This innovative approach can assist farmers in acquiring improved breeds of goats. Other farmers in the community may also buy the improved goats from those with access to a buck.

8.3.3 Strengthen Linking Social Capital

Linking social capital is vital in providing safety nets for smallholder farming households during shocks and stresses. Mwakiwa et al. observe that linking capital influences adaptive capacity, sensitivity and disaster exposure. With linking capital, households can ride the rich network with stakeholders in their wards or district or even outside of their district and get access to important information for enhancing resilience to shocks. The information is critical for decision-making during disasters such as droughts and social shocks such as COVID-19. Zamasiya, Nyikahadzoi and Mataruse reiterated that linking capital is critical for strengthening resilience to shocks and stresses and helps smallholder farmers avoid using food-based and asset-based livelihood coping strategies during shocks. By establishing activities that comprised members from different villages and linking them with strategic stakeholders, the ZRBF strengthened linking capital. It is, therefore, noted that project interventions should deliberately target the strengthening of linking capital to enhance resilience to shocks and stresses.

Linking social capital can be enhanced by deliberately designing activities that cause smallholder farmers to collaborate and work with other stakeholders such as NGOs and government workers. For instance, a project can introduce the rehabilitation or building of a dip tank in a community. This will cause the community to unite and link up with various stakeholders, such as traditional leaders, rural district councils, NGO workers and Livestock Production Department Officers. The activities around the dip tank can enhance resilience to shocks by improving access to information on dipping chemicals, cattle diseases, cattle movements and market dynamics.

8.3.4 Promote Access to Innovative Financial Services

Access to financial services is essential in enhancing farming households' resilience to shocks and stresses. Most smallholder farmers fail to access financial resources due to a lack of collateral and poor access to urban centres. This results in them failing to access much-needed financial resources to boost their enterprises. Zamasiya, Nyikahadzoi and Mataruse noted that smallholder farmers with access to financial services, either as loans or savings, can better adapt to shocks and stresses. The ZRBF programme created internal savings and lending schemes (ISALs), facilitating the culture of savings among smallholder farmers. Improvement of access to financial services reduces the use of negative coping strategies. Mwakiwa et al. noted that although preferable, credit or purchase on credit are usually unavailable during stressing periods. To enhance smallholder farmers' resilience to shocks and stresses, it is vital that resilience-building initiatives deliberately ensure that smallholder farmers have access to financial resources.

Access to financial services can be enhanced by organising smallholder farmers into internal savings and lending schemes (ISALs). The ISALs can help to mobilise savings among smallholder farmers and provide credit for others to expand their businesses or working capital requirements. The initial capital for the ISAL can either be mobilised by members or injected by the project as seed capital/revolving fund/matching grant. The smallholder farmers can then be linked to private produce buyers so that they produce for the market and generate financial resources.

8.3.5 Facilitate Accumulation of Productive Asset Base

Zamasiya, Nyikahadzoi and Mataruse indicated that productive asset accumulation makes farmers to become better prepared to manage climate induced shocks. The role of productive assets is to enhance access to information for timely decision-making and provide the means for the smallholder farmers to implement a decision. In the absence of productive assets such as phones, smallholder farmers struggle to access timely information, such as early warning systems or market prices for their produce. Productive assets allow smallholder farmers to take advantage of opportunities. For instance, ownership of an oxen and a plough may mean farming households can capitalise on opportunistic rainfall and plant their crops on time. Mwakiwa et al. indicated that productive assets are significant for strengthening the resilience of smallholder farmers. However, they note that poorer households tend to stress sell their productive assets during shocks and stresses, leading to worsening resilience. It can, therefore, be noted that resilience building should ensure that smallholder farmers can build productive assets.

Development practitioners must introduce integrated or value-chain programming to smallholders to enhance the accumulation of productive assets. An innovative way could involve targeting cattle farmers with activities on producing cattle feed (fodder or hay from local materials or stover preservation). The fodder and hay can be used as supplementary feed and help cattle farmers avoid stress selling or destocking during climate-induced stresses. The cattle farmers can also be introduced to disease management, where they must use spray races instead of dip tanks. The use of spray races helps the smallholder farmers to control tick-borne diseases on their cattle, especially where dip tank services may be difficult to access due to shortage of chemicals or dry conditions. Spray races will help cattle farmers avoid stress selling to reduce tick-induced deaths. Productive assets can also be enhanced by exposing smallholder farmers to diversified income opportunities within a project. If an environmental shock adversely affects one of the income streams, the smallholder farmers can rely on other income options. Diversified income streams improve smallholder farmers' income and reduce their propensity to stress sell productive assets.

8.3.6 Promote Human Capital Development

Human capital is pivotal in enhancing smallholder farmers' adaptation to climate change. Mwakiwa et al. and Zamasiya, Nyikahadzoi and Mataruse highlighted that the ZRBF programme improved the human capital of smallholders, resulting in them adapting better. Activities such as workshops, crop-livestock information centres and exchange visits are more likely to assist farmers to understand livestock disease calendars. They can then use their accumulated human capital to make informed decisions on how to adapt to climate change. Mwakiwa et al. noted that households with higher levels of human capital are better able to make decisions on adaptation, leaving them less vulnerable to shocks and stresses. Smallholder farmers with higher human capital can better understand crop and animal diseases. This makes them use the information and address the shock better. Resilience-building initiatives should deliberately enhance human capital development by layering and sequencing programme interventions. Layering can be done by exposing smallholder farmers participating in different project activities or within activity types. The exposure can be facilitated through Crop and Livestock Information Centres where smallholder farmers can learn the various components. Through this approach, smallholder farmers are more likely to learn and enhance their knowledge and be able to make appropriate decisions to enhance their resilience.

8.3.7 Enhance the Provision of Water Infrastructure

Access to water infrastructure for women and youth is one of the critical factors that can help improve smallholder farmers' resilience to environmental shocks and stresses. Bhatasara and Karuma demonstrated that providing water infrastructure close to communities' homes frees up time for women to embark on other income-generating activities. In most cases, droughts result in the drying up of other water sources, such as rivers and shallow wells. These sources usually provide water to households for various household uses and gardening. Because of their proximity to communities, women and youth spend less time fetching water. However, shocks and stresses considerably constrain the women and youth who need to walk longer distances to fetch water. The ZRBF programme provided water infrastructure close to the communities. This provision ensured that women were spending less time on fetching water. It is, therefore, essential to note that providing water infrastructure close to communities improves their resilience to climate-induced shocks.

8.3.8 Target Young and Poor Households for Resilience-Building

Vulnerability to environmental, social and economic shocks and stresses is a function of household characteristics (age, gender of household head, etc.). Mwakiwa et al. noted that young smallholder farmers are highly vulnerable to shocks and stresses. This is mainly because they lack experience and ownership of strategic asset management for shocks. Manjengwa et al. observed that the poor are the most vulnerable to shocks and stresses. In the event of prolonged shocks and stresses, the poor quickly become poorer and have a high propensity to resort to negative coping strategies. Nyika et al. observed that younger farmers are more likely to adopt CSA technologies than older farmers. When presented with the right resilience-building opportunities, young people can improve their adaptive and absorptive capabilities and better withstand shocks and stresses. This generally points out that resilience programmes should target the younger and poor smallholder farmers as these are more vulnerable to shocks and stresses. The young and poor farmers can be organised into groups of ten or fewer farmers. For sustainability, groups with young and poor farmers should be fused with older, resource-endowed farmers. This will assist in stabilising the groups as the youth are very mobile, which may lead to the collapse of groups.

Index

L. Mhlanga et al. (eds.), *Building Resilience to Climate Change in Selected Districts in Zimbabwe*, https://doi.org/10.1007/978-3-031-80529-5

The manufacturer's authorised representative in the EU is Springer Nature Customer Service Centre GmbH, Europaplatz 3, 69115 Heidelberg, Germany. If you have any concerns regarding our products, please contact ProductSafety@springernature.com

Printed and bound by CPI Group (UK) Ltd, Croydon, CR0 4YY
01/10/2026
02225945-0012